Optimistic Storm: Part 3: A Future Powered by Ingenuity

Hydrogen, Nuclear and Other Green Tech Fails

> *Specialists are people who always repeat the same mistakes.*
>
> 1883-1969 Walter Gropius.

Dr Hugh Butler

ISBN: 978-0-9941733-6-2 printed
ISB: 978-0-9941733-7-9 e-book

Part 3: Nuclear and hydrogen technologies at best distract us and at worst increase the severity of climate change and worsen impacts on people, the economy and our environment.

Structure of Book

These books and chapters are interconnected and we cannot talk about one without referencing multiple other subjects. Electricity is foundational. For the first time since humans walked on the earth, we no longer need to burn molecules for energy.

- Part 1 explores electricity disruption and the end of the fossil fuel era. The electrification of all energy cascades into transportation and stops the emissions that cause climate change. The electrification of energy changes our society.
- Part 2 explains how AI is about to transform food, health, agriculture, transport and labour.
- Part 3 concludes nuclear, hydrogen, carbon capture and storage and direct air capture distract us from our energy and AI-enabled future.

The rise of free energy and free labour will transform our societies and our home, Earth.

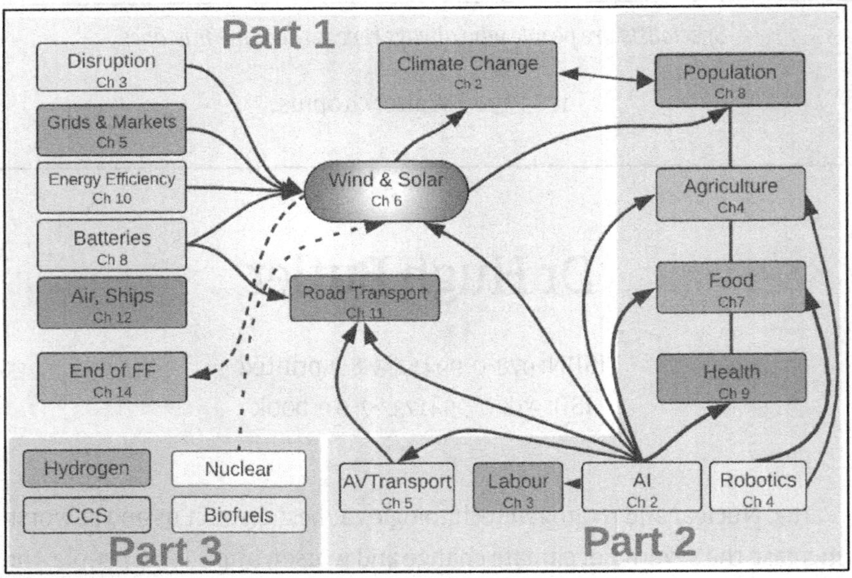

Figure 1 Outline of Optimistic Storm books

This v1.1 edition includes feedback and comments for clarity.

Table of Contents

1 Introduction

Explore the Chapter:

- The climate crisis demands we reduce emissions. (Part 1)
- The "carrying capacity" of the world (Part 2) drives urgency to expand clean electricity and reduce the destruction our environment.
- We will summarise some topics that are in Part 1 and Part 2 of Optimistic Storm.
- How a misunderstanding of the technology and economics of technologies, including hydrogen, nuclear, carbon capture and storage and direct air capture, has led businesses and politicians down blind economic paths.
- Explore the policy failures and Political murkiness behind these technologies.

In Part 1, we explored how we are in a climate crisis. Each year brings new records, new heat waves, and new anxieties about the future. Headlines scream of impending doom, painting a picture of a world ravaged by climate change, where rising seas swallow coastlines and extreme weather events become commonplace. It's enough to feel the weight of despair pressing down, a sense of helplessness in the face of such a monumental challenge. The best estimate is that for every 1°C of increased global temperature, GDP will take a 12% fall. Initial estimates in 1997 were 2% per degree, but the economic models say that at 3°C we will suffer a fall of GDP of 36% -[1] just like living in the Great Depression, but forever. [2] This is preventable. We we have the technology and economics to stop going to that scenario. Politics can either slow that transition down or accelerate it.

We can be very optimistic as we explore the end of the fossil fuel era. For the last million years, our primary energy source was our muscles (Era 1). The invention of coal-powered steam engines marked the beginning of the Industrial Revolution. A hundred years later, steam-powered rail and shipping opened up the world. (Era 2). Gas and oil took us through Era 3. We are now entering Era 4 of energy production, where we harvest electrons. This marks the beginning of the end of burning molecules. Despite claims of unreliability, electrons from wind, solar, and batteries are cheaper, more efficient, and safer than burning molecules for energy. It ends environmental destruction and waste caused by fossil fuel burning and nuclear energy. We transition from consuming 500 years' worth of stored carbon to using just 1.5 hours of sunlight. We can see what we need to do to replace 166,000 TWh of fossil fuel with just 66,000 TWh of electricity. How personal public transport using electric vehicles will end the era of 1.4 billion internal-combustion-engine vehicles. Moreover, we can see a pathway to end energy poverty for 800 million people. While this marks the end of the fossil fuel era, we are left with enormous heat stored in the oceans and high levels of greenhouse gases in the atmosphere. These will continue to exacerbate global warming and land degradation, as well as food, health, and societal issues.

In Part 2 of The Optimistic Storm, we explored a range of instances where AI will:

- Augment or replace knowledge workers. These white-collar workers are over 80% of those employed in Western countries.
- Empower humanoid robots that will replace most blue-collar workers. The other 20%. Within 2 decades (2045) I predict there will not be one job that cannot be done by a machine.
- Advance science leading to better health, more drugs and bring health to over 6.5 billion people globally with poor health services.

- Enable precision fermentation to reduce the feeding of 40% of our food to feed animals.
- Enable the production of protein at 30 times less resource inputs such as land, fertilisers, pesticides and energy.

We also looked at the world's carrying capacity – birth rates, life expectancy, and projections of how many resources we will need for an electric energy and transport system. We also showed that these advances will shrink the land required for agriculture by an area equal to the combined land area of the USA, Australia, and China.

This Part 3 book deals with some technologies for the zero-emissions renewable electricity that we did not discuss in Part 1. Nuclear, hydrogen, Carbon Capture and Storage, Direct Air Capture and biofuels are not part of the Optimist Storm story. They are part of the past era of molecular energy. Further, they are significant distractions and impediments to achieving zero emissions, reducing pollution, and addressing societal woes, which we will examine in more detail. The reason they still get airtime is purely economic decisions by the fossil fuel industry. That industry is pushing political and personal belief systems. We will explain why well-meaning people and countries continue to promote or invest in these technologies, directly contradicting both the technology and the economics, which prioritise the needs of people, not corporations.

1.1 Why People Change

Dr Adam Dorr of RethinkX in his book Brighter: Optimism, Progress, and the Future of Environmentalism, [3] laid out the premise that prosperity will help drive change. A business colleague once told me people only change for one of three drivers of change.

- They make money
- They lose money
- It's the law

As we will discover here, these molecular technologies continue to be profitable but only because of the $14 trillion (and rising) in taxpayer subsidies. When those subsidies run out, those stranded assets will be no more. Worse, that money could have accelerated the transition from molecules to electrons. Instead, we have more emissions and delay progress to zero emissions.

1.2 Phase Change Disruptions

For many types of innovation, the new product or service coexists with existing ones. It is not always the case. We explored a particular type of innovation called "Phase-Change Disruption", discussed in detail in Part 1, Chapter 3. As Tony Seba points out, these Phase-change disruptions create entirely new, significantly larger systems with distinct properties, novel business models, and distinct metrics. We have this phase change disruption occurring in:

- Energy
- Transportation
- Labour
- Health
- Food and Agriculture

It is impossible to know in advance the full details of how these systems will differ from those we have today. With AI, we are past the rupture point, but are we on the S-Curve Adoption, or are we before that?

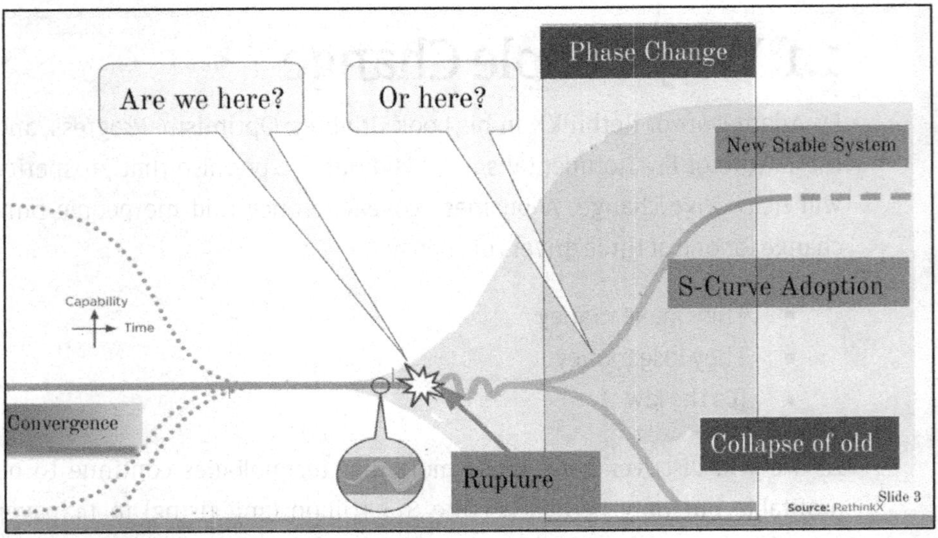

Figure 2 Phase Change disruption as described by Tony Seba (RethinkX)

Figure 2 shows that a phase change disruption needs range of technologies to converge. Sometimes these technologies do not appear to have much in common, yet combined they can lead to a rupture point. For example, the portable music industry needed good batteries. No one in the 1980s would have envisaged that these batteries, connected together, would be the basis of electric vehicles. A key feature of many new technologies is that the marginal cost for both renewable energy and labour will fall and rapidly approach zero. As they do, other technologies will become enabled because most of our wealth and cost of goods and services is from raw materials, energy and labour. All goods and services, including food, housing and transport, will become cheaper.

1.3 Technology Enables Disruptions

We also explored that these changes are complex system changes. Ark Invest CEO Cathie Wood talks about 14 technology disruptions as core to the five societal disruptions. Tony Seba of RethinkX says three societal disruptions need eight technologies. Neither Ark nor RethinkX see nuclear or hydrogen as part of those technologies, and we will show in the following chapters that the barriers include:

- No clear technology changes during the past decades.
- No compelling economic advantage.
- They have to compete with other disruptive technologies.
- No long-term sustainability advantages and face increasing competition from other technologies.
- Negative learning curves. They become more expensive. Not cheaper.

In other words, they have run their course, with nothing new or substantive and moreover, the gap with other technologies widens over time. Any commentary that says they are about to enter a new renaissance is false. These molecular technologies have reached a dead end. This is not to say they will not have niche markets, but they will survive only with substantial political interference and subsidies.

1.4 Political Murkiness

Information is often distorted by commercial interests. Climate obstruction in different industries is more coordinated than is generally recognised. We have evidence of "coordinated climate obstruction efforts" among oil, plastics, and agrichemical industries in social media messaging. In Figure 3, we see the climate denial networks at work.

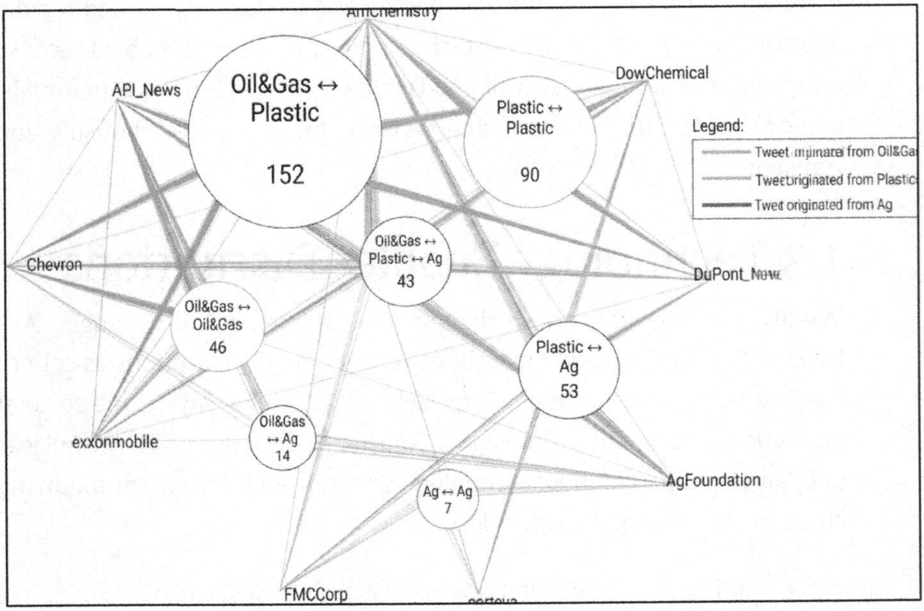

Figure 3 Networks of climate obstruction by Kinol et al 2025.

In Figure 3, DeSmog [4] shows from 2008 to 2023, nine of the USA largest oil, agrichemical, and plastics trade groups and corporations posted thousands of tweets on the social media platform X, and their messaging on environmental issues was strikingly "obstructive" for climate policy and action. [5] A Venn diagram of fossil fuel companies (Chevron, ExxonMobil), agrichemical companies (FMC Corporation, Corteva, Ag Foundation), and plastics (Dow, DuPont, American Chemistry) all came together at the intersection: petrochemicals, with the number of tweets over 152. With AI tools and social media sites, the noise can be deafening.

We are seeing similar increases in media releases for hydrogen, nuclear, carbon capture and storage and direct air capture.

1.5 The ElectroTech Revolution

The consulting company Ember says we are in the "Electrotech" revolution, [6] although Optimistic Storm adapts Prof. Vaclav Sinil's definition [7] of energy eras. In this 4th era of energy, many respond with disbelief and resistance, as did horse owners from the 1800s to the 1900s. We've relied on energy from muscles (Era 1) and heat energy from rocks (also molecular energy) to help us do work (Eras 2 and 3) and are now moving to the essentially unlimited and free energy from the sun, harvesting electrons to provide much of what is needed for life (Era 4).

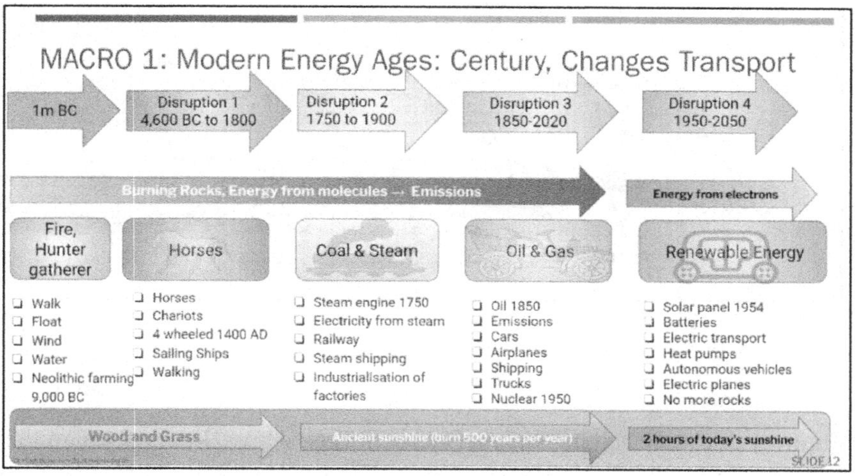

Figure 4 The 4 eras of energy and transport changing form molecules to electrons

These Eras in Figure 4 over the past 7,000 years show the energy transition since we first used coal for energy, which took over 100 years to reach maturity. Transport followed the change in energy, but followed the change in energy by about 100 years. If we review how far we are through Era 4 on the phase change disruption image (Figure 2) it is probably less than 5%. AI is less than 1%. For all the changes we see today, nothing will be like what it will be within two decades. Moreover, this period of transition is very tumultuous, with the new coming on stream while the old systems appear to be clinging on. While this is a Phase-Change disruption event, many get confused or doubt the change is happening, because the existing fossil fuel use is not yet decreasing. Meanwhile the new system is providing all the additional energy demand. When will the existing system start to decline?

Most data suggest we are on the cusp today, but we may need another 3 or more years to see fossil fuel energy peak and begin to decline.

Reports of my death are greatly exaggerated.
Mark Twain

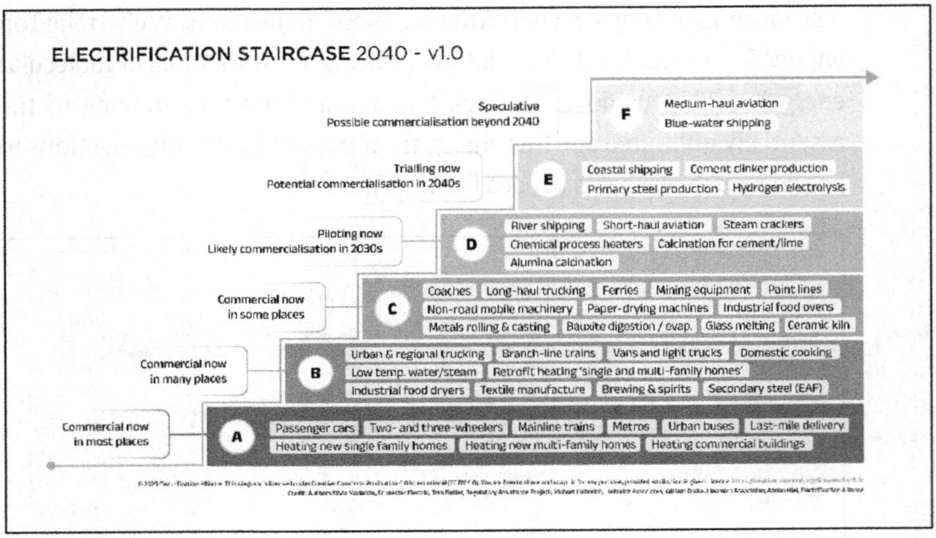

Figure 5: Electrification Staircase by Michael Liebreich, showing the status of sectors in electrification

Figure 5 shows this change in energy in various sectors. Michael Liebreitch describes this as an electrification staircase with staircase showing at what stage different sectors are at. Levels A, B and C sectors show fossil fuel is out, and electrification is commercial and available now. Other sectors (Level D, E, F) may take 2 to 3 decades, if at all.

1.6 Speed of Change

If we look at the last seven years of change, it would have been impossible to predict that we had a global pandemic, a war in Europe and that machines would be able to talk and nearly be as smart as humans. We have also seen a radical change in energy.

- Clean energy is now the default source of growth for electricity supply.
- Economics are improving with costs collapsing.

- Capital has moved decisively. Investment has stopped going into fossil fuel industries and is rising into new clean green electricity.
- Electric energy logic. Decarbonise, electrify everything, everywhere.
- Constraints are no longer infrastructure or technical, but just delivery problems.
- The politics around climate change have become more volatile as the fossil fuel industry tries to slow down the transition.
- Public awareness has increased, but is neither aligned nor linear, which means there has been an ineffective political outcome.

Where do nuclear, carbon capture and storage, biofuel and direct air capture fit with this global change in energy supply and demand? Mark Twain died 13 years after he made his quote in 1897, and it reminds us of similar proclamations by fossil fuel and nuclear companies. Do they have 13 years of life left? Maybe more, maybe less. Certainly not decades.

Conclusion of Introduction:

- Phase change disruption is underway in the energy system.
- The 4th Era marks the changes from molecular energy tc electron harvesting.
- Clean renewable electricity is continuing to fall in cost.
- There are companies looking to use their political anc organisations wishing to slow down or stop this era change.
- Some technologies are part of this change. Hydrogen, nuclear carbon capture and storage, biofuel, e-fuel and direct ai: capture are not likely to be – as explored in subsequen chapters.

2 Hydrogen

Exploring Hydrogen:

- What is Hydrogen (H_2)?
- Technology and economics?
- Existing uses and potential uses?
- Why was H_2 promoted extensively?
- Use the "Hydrogen Ladder" to frame the narrative.
- Explore future uses for H_2 in transport, heating and storage, steel making and e-fuels.

Hydrogen (H) is in the news these days – all the great things it will do. [8] H_2 provides an interesting insight into how a product claimed to end the fossil fuel era is, in fact, just a very limited technology. We do need a solution to replace the H_2 gas made from fossil fuels, and we do need to reduce emissions in some uses of H_2, but the reductions are much smaller than proponents have argued. Some forecasts hydrogen to contribute 6% of total global emissions reductions by 2050 [9], positioning it as a key enabler in the global decarbonization effort. Others are less optimistic.

Before we position hydrogen as the solution to climate change, we first have to deal with hydrogen as a problem in climate change.
Michael Liebreight *[10]*

Back in 1875, Jules Verne said, "*Water will one day fuel; that hydrogen and oxygen which constitute it, used singly or together, will furnish an inexhaustible source of heat and light, of an intensity that coal is not capable.*" He should have checked with Sadi Carnot from 1824 or Clausius and Thompson from the 1850s Laws of Thermodynamics. But it was Jeremy Rifkin, with his The Hydrogen Economy in 2003, who created the fantasy of the worldwide energy web. In reality, we are living with 20 years of hydrogen myths spread by reporters with a poor understanding of science and technology and hoodwinked by fossil fuel companies. Hydrogen is the smallest and lightest element, but it has some interesting properties when compressed under high pressure and heat. There are even sound theories that the inside of the Earth includes 50 times more hydrogen at 50 Gpa and 3,500°C than in the oceans. [11] But that is another story. A key understanding is that hydrogen can be used either as an energy store or as a chemical feedstock or precursor to other compounds. It is not a source of energy except for the limited reserves of naturally occurring H_2 gas that could be harvested as an energy source.

2.1 H_2 Technology

Hydrogen has been produced since 1650, when Théodore de Mayerne first poured dilute sulfuric acid on iron to produce a gas of "inflammable air". A hundred years later, in 1783, Jacques Charles made a hydrogen balloon large enough to carry him and a colleague over a distance of 36 km to a height of up to 550 m, demonstrating that hydrogen had other uses. Three subsequent discoveries opened up the possibility of realising its chemical potential. These were hydrogenation (1897), the Haber process to make ammonia (1910), and hydrocracking (1920).

2.1.1 Source or Colours of H_2

Hydrogen is often described in colours, although these colours are arbitrary, as physically the H_2 molecule has no colour and hydrogen is hydrogen. Still, the colour is used to reflect the ways hydrogen is sourced – through steam reforming of fossil fuels, through electrolysis of water to produce CO2 and H2, and through mining.

- Grey H_2 - from gas.
- Black H_2 - from lignite (a small amount these days).

- Blue H_2 – made from gas, and then CO_2 is captured using CCS.
- Green H_2– made from water and renewable electricity.
- Pink H_2- like green, but using electricity made by nuclear.
- White H_2- mined from natural geological H_2 deposits.

2.1.2 Grey and Blue H_2 Production

The majority of production today uses the Haber process - or the steam methane reforming (SMR) process with methane as the feedstock, and non-renewable energy powering production. Methane is split into H_2 and CO_2, and the H_2 is called grey hydrogen. If CCS technology captures the CO_2 emitted during production, the process is considered blue. We explore carbon capture in the CCS chapter and show how bad the economics are and the failures, both technically and economically. There is every reason to suggest that blue H_2 using CCS will also be an economic failure. Blue hydrogen is only blue if the carbon capture and storage (CCS) system captures the CO_2 emissions effectively and efficiently over the long term. It is misleading to call it "blue" if the CCS system addresses only a small fraction of the project's emissions. [12] A major source of greenhouse gas emissions is from gas wells. The average US gas well has over 1.5% leakage during extraction. Recent aerial studies suggest that leakage may be as high as 9% in oil-heavy basins. At average US methane leakage rates, producing hydrogen from natural gas via steam methane reforming will have cradle-to-gate emissions above 2.5 kg CO_2e/kg H_2, even if producers are somehow able to capture 100% of CO_2 emitted during production.[12]

2.1.3 Green H_2 Production

Hydrogen can also be produced from water with renewable energy – this is what's known as "green hydrogen." Currently, it makes up less than 1% globally.

The US Office of Energy Efficiency and Renewable Energy has a "moonshot" for H_2 which is three "ones".

- Reduce cost by 80% to "one" dollar.
- Per "one" kg.
- in "one" decade.

They claimed this would lead to a fivefold increase in clean hydrogen use.[13] When you look at what the process is to make green H2, only a fraction of the challenge and cost is the actual chemical reaction.

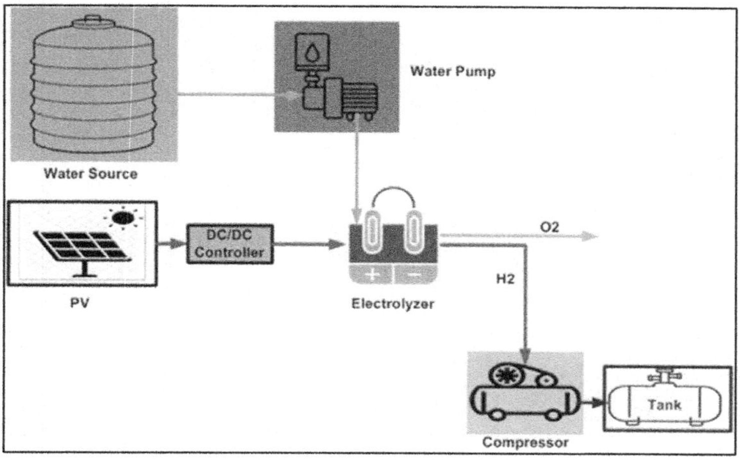

Figure 6 Schematic of green hydrogen production process

Figure 6 shows schematically what is needed. A water source. Clean green zero emission renewable electricity and a stream of CO_2 in a highly complex and expensive electrolyser. Storage and transport is needed to where it is used.

2.1.4 Green H_2 Prediction Versus Reality

Carbon Brief highlights the gap between prediction versus reality for green H_2. In Figure 7 shows hydrogen projections from oil companies, NGOs and academics from 1970, running up to 2070, against actual historical production (million tonnes per year). It shows how out of touch they were. [14] Most made the mistake that the process could rapidly scale and come down in cost and they used the changes from the solar industry to say costs would fall. However, green hydrogen is not just the electrolyser where the water is split. The downstream processes are that the H_2 is compressed, stored and transported.[15]. The mechanical parts of a system (pumps and storage) are from a mature industry. Apart from electrolysers, there are n readily available cost-down technologies available because the existing gray hydrogen industry is mature. They make about 100 million tonnes per year so any new innovation would have been invented and adopted to bring the cost of grey hydrogen down.

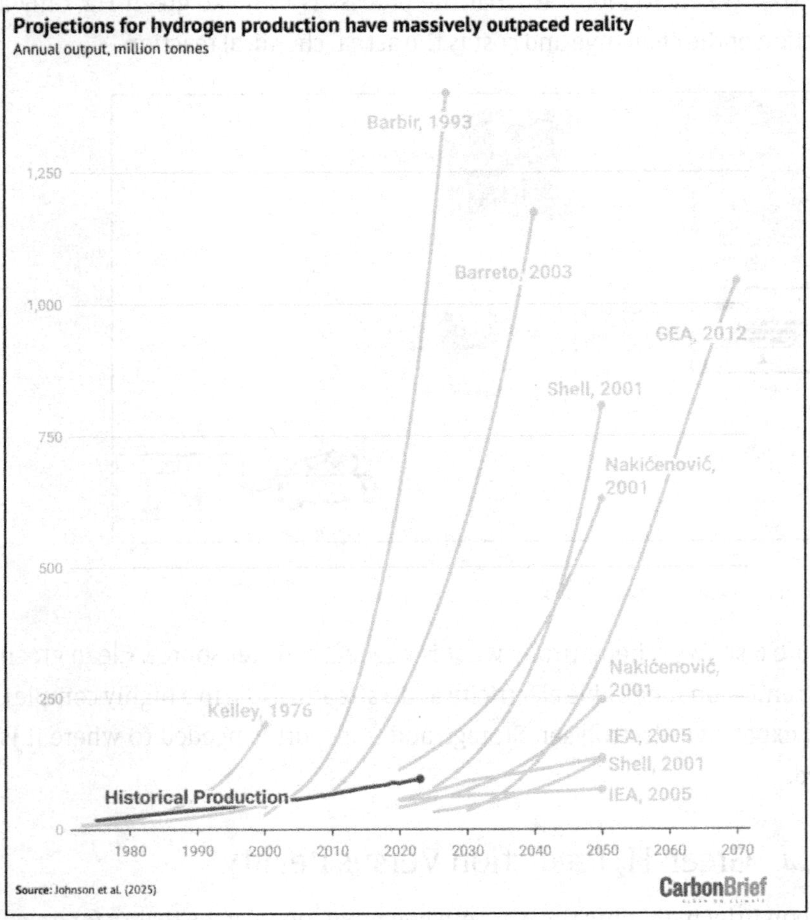

Figure 7 Historical production versus prediction by academics and fossil fuel companies since 1980 to current

Figure 7 is the complete opposite of projections by incumbent consulting companies about the rise of battery and solar. Those organisations thought that by 2030, renewable energy might account for a few per cent of electricity generation, and that EV sales would be less than 10% by 2035.

2.1.5 Geologic or White H_2

Geologic hydrogen, variously called white, gold, natural or orange hydrogen naturally occurs underground. White H_2 could be a clean, cheap source of energy. [16] No one actually knows how geological hydrogen is produced

[17] although there are three methods.

- Serpentinization. Probably the most significant process. [18] It occurs when water reacts with iron-rich rocks like olivine under high temperatures and pressures. It involves hydrolysis and transformation of primary ferromagnesian minerals such as olivine ($(Mg,Fe)_2SiO_4$) and pyroxenes ($(Mg,Fe)SiO_3$) to produce H_2-rich fluids. Essentially, the iron in the rock oxidises, releasing hydrogen as a by-product.
- Radiolysis. Radioactive decay of elements like uranium and thorium can also produce hydrogen by splitting water molecules.
- Mantle Degassing. This theory suggests that hydrogen trapped within the Earth's mantle can be released through tectonic activity.

Extraction remains in the early stages of development, with only one site in Mali and facing various technical and economic challenges. There are three main methods to extract geologic hydrogen: from traps, directly from underground rock formations, and through stimulated hydrogen production (similar to fracking). The current hotspot areas for geologic hydrogen include North America, Australia, and parts of Eastern Europe and Northern Asia. Prospectors are rushing to find it – drilling for hydrogen in northeast France, Australia, Spain, Morocco, Brazil, and in the United States, in Nebraska, Arizona, and Kansas. Australian exploration company HyTerra discovered H_2 concentrations of 83% at its latest exploratory well and previously found 96% concentrations at Kansas at depths of 1,000 to 1,700 m according to Aug 2025 market announcements. [19] They have a way to go, as they need to set up production test wells prior to production.

Co-producing helium with geologic hydrogen could improve its economics. Helium is in short supply, with a global market of 170,000 m^3 of bulk liquid helium valued at over $2.8 billion, with Qatar and the USA as major producers. It is currently co-produced from gas extraction. There is debate about the potential of geologic hydrogen. Optimists believe it could be the cheapest and cleanest form of hydrogen, while sceptics highlight the lack of large-scale production and transportation challenges. Fugitive emissions are a concern from extraction, storage and transport. Hydrogen leaks from wellheads, pipelines, and other infrastructure are inevitable. Emissions will happen regardless of how hydrogen is produced. Naturally or manufactured – it leaks. The LCA is expected to be about 0.4kg CO_2/kg. The estimated cost would be about $1/kg, and hence the gold rush. With green hydrogen looking as if it will disappear will not be a "positive driver" for geologic hydrogen. It'll probably be a death sentence. Those who promote geologic H_2 say investors and policymakers need to understand the difference between them and the advantages of geologic H_2. The reality? H_2 is H_2 no matter the source. It is just dollars.

2.2 Economics of H_2

The conventional grey hydrogen benchmark price is about $2 per kg. Using a long-term average of $75/bbl for oil and $4-6/GJ for natural gas, then green H_2 vs black hydrogen is uncompetitive.

Type	Process	Carbon Intensity kgCO_2/kg	Production $/kg
Green	Electrolysis	Less than 1	8-12
Grey – fossil fuel	Reforming	10-14	2-6
Brown – fossil fuel	Gasification	19-26	2-3
Blue	Grey + CCS	1.7- 8.6	4-8
Geological	Mining	0.4	?

Figure 8 Carbon Intensity of various Hydrogen production methods

It is estimated that at 56% CO_2 capture, blue hydrogen's carbon intensity can be reduced to around 6.87 kg CO_2/kg. At 90% capture rate, it might be further reduced to 4kg CO_2/kg. [20] Other studies show that blue hydrogen may be about 10% better than grey if the full lifecycle is assessed.[21]

- Cost of a hydrogen electrolyser has fallen with new announcements from multiple companies and is reduced to about $300kW (or $3 million per MW). Some companies have gone bust.

- Plant life cycle is 10 years. (electrolysers have limited life)
- Modelled hydrogen cost is often stated to be about $4 to $5 per kg, but the Netherlands' most recent is over €12 ($13). [22]
- The energy from 1kg of hydrogen is 33 kWh
- Production cost = two to three times more expensive
- Pipelines = 10-50% more expensive
- Fuel cells and storage tanks for road transport = 2 to 3 times more expensive than internal combustion engines.
- Synthetic fuels for aviation = 3-6 times more expensive than Jet A1.
- Ammonia = 50-75% more expensive.
- Methanol = 150%.
- Steel = 30-40%.

These costs are not a pathway to success. According to a recent report by Dutch research institute TNO, funded by the Ministry of Economic Affairs and Climate Policy, based on actual project data, the average levelised cost of large-scale production facilities built today would be more than €13 ($14.15)/kg H_2. [23]. The $1 per kg is at best disingenuous. Even when we look at white (gold) hydrogen, the estimated costs are uncertain.

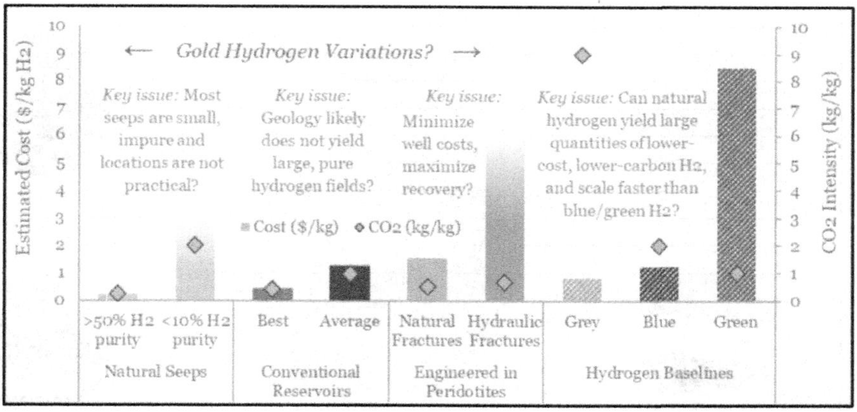

Figure 9 Costs and Carbon Intensity of gold H2 versus various other H2 colours.

In Figure 9, the columns are the expected cost of the H2, and the diamonds are the CO2 intensive values.

2.3 Current Uses of H_2

Today, the world makes 94 million tonnes of grey and black hydrogen from unabated natural gas and coal. It emits 830 million tonnes of carbon, and these figures continue to rise. Those emissions are more than 2% of global CO2 emissions, comparable to the combined annual emissions of Indonesia and the United Kingdom.

2.3.1 Use in Fossil Fuel Refining

Over 25% of hydrogen use is in oil refining, used to remove chemicals such as sulphur from the crude oil. Sulphur is a highly versatile chemical element primarily used to produce sulfuric acid, which is essential for manufacturing fertilisers, batteries, and chemicals. Beyond industrial chemistry, it is used in agriculture as a fungicide, in rubber vulcanisation to harden tyres, and in dermatology to treat acne and dandruff. With a fall in oil demand with electrification of transport (65% of oil produced) the demand for H_2 will fall in the coming decades, unless there is some other use case.

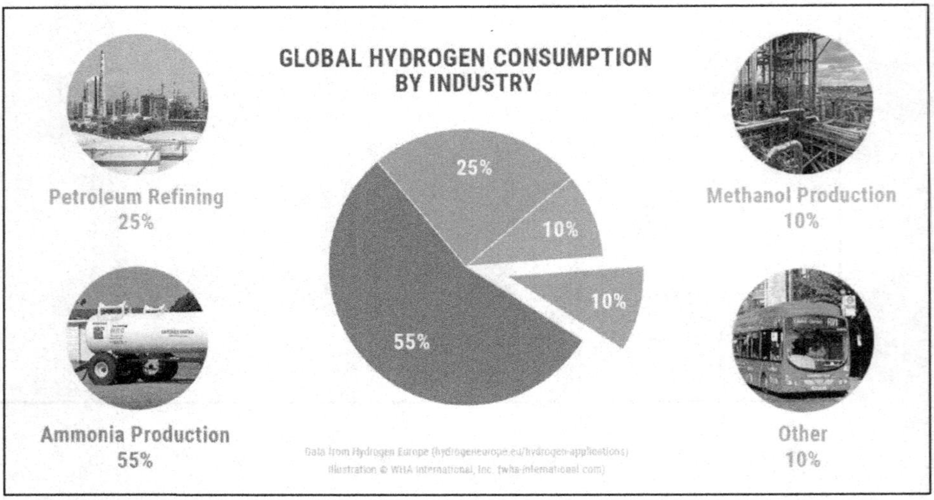

Figure 10 Use of hydrogen globally for refining

Figure 10 shows that approximately 55% of the world's hydrogen is used for ammonia production, 25% for refining, and around 10% for methanol production. Other uses make up less than 1%.

2.3.2 Ammonia Production

Most synthetic nitrogen fertilisers (such as urea, ammonium nitrate, and UAN) are created using the Haber-Bosch process. This process uses methane (from natural gas) as both a fuel and a raw material to extract hydrogen. That hydrogen is then combined with nitrogen from the air to create ammonia, the base for nearly all nitrogen fertilisers. Nitrogen fertilisers help increase crop and pasture yields. We will explore this in more detail in the fertiliser uses in a later section.

2.3.3 Food Hydrogenation of Fats

One minor use of H_2 is in hydrogenation, which converts unsaturated fats in vegetable oils into saturated fats. This process creates fats that are more solid at room temperature, which is desirable for some food applications, including the production of margarine, shortening, and cooking oils. A discussion around the rights and wrongs of unsaturated versus saturated fats was detailed in my book The TakeOut Diet a decade ago. Dieticians and doctors are still arguing.

Precision fermentation will disrupt this whole discussion about hydrogenation. We currently grow crops, corn, cottonseed, safflower, soy, and sunflower, from which we harvest the seeds and extract oil. Precision fermentation bypasses all the growing and manufacturing. Sunflowers need fertiliser and 200 litres of water for 1kg. For more information on the Precision Fermentation of oils, check out Part 2 of Optimistic Storm. That the uses of H_2 will be disrupted from outside the gas industry would come as a surprise to some, but it is common in disruptions.

2.3.4 Manufacturing Pharmaceutical Products

Hydrogen is used as a reactant or solvent in the synthesis of various vitamins and other pharmaceutical drugs. For example, H_2 is used in the production of vitamins A and E and steroids. All these are available in natural food products, Precision fermentation can already produce these albeit at an increased cost, but that cost is rapidly falling. Other applications, such as food, metalworking, welding, glass production, electronics, and renewable energy, account for less than 1%. Most production uses gas piped to the site, converted to H_2 and used on site. [24] Manufacturing hydrogen today creates as much CO_2 as all global aviation in 2019, so it is an important sector to decarbonise eventually. It is just a matter of when.

2.4 Ugly Features of H_2

Like many aspects of our modern society, H_2 has some nasty properties due to its size and reactivity.

2.4.1 H_2 is A Bad Greenhouse Gas

A primary concern about H_2 is that it is an indirect greenhouse gas. It is not classified as a direct greenhouse gas under the definitions of the Intergovernmental Panel on Climate Change (IPCC). [25] While not a direct global warming gas, hydrogen is recognised as an indirect climate forcer, contributing to global warming through the mechanism described below.

It inhibits the breakdown of methane. As we discussed in Part 1, methane is a major contributor to global emissions, from coal and gas leakages and directly from agriculture.

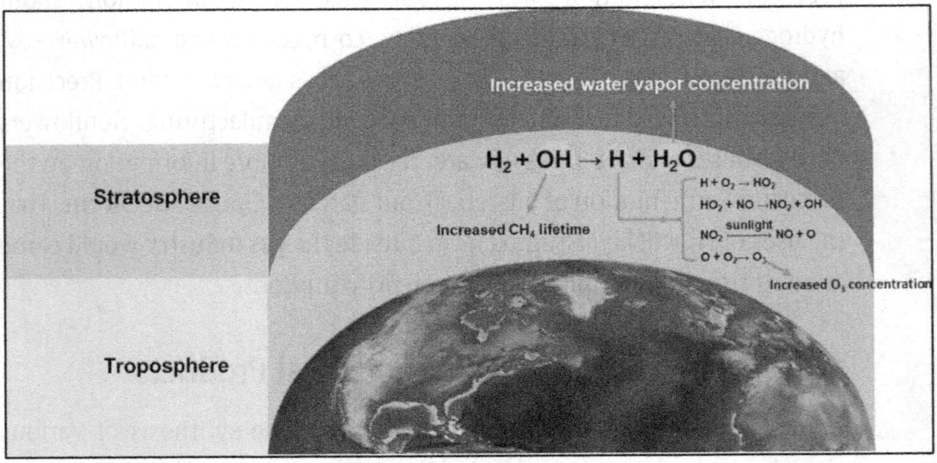

Figure 11 Global warming impacts of hydrogen in troposphere by Rezaei et al 2025

Figure 11 shows the pathways of H2 for increasing climate warming. In context, the warming potential of methane is usually expressed as CO_2 equivalent. Methane has a high global warming potential (GWP100) of 28, making it 28 times more potent than CO_2. There is another kicker. CO_2 takes about 100 years to break down. Methane takes about 20 years. Therefore, methane has a GWP20 of 86, meaning that in 20 years it contributes about 86 times as much as CO_2. Every effort needs to go into preventing methane emissions today to prevent effects on climate change in 20 years.

The chemistry is pretty straightforward. There are hydroxyl radicals – one oxygen and one hydrogen joined together in a highly reactive pair – in the atmosphere that the gas methane interacts with and causes its breakdown into carbon dioxide and water vapour. Hydrogen is also very reactive, and when it's in the atmosphere, it competes very effectively for the relatively limited supply of hydroxyl radicals. That means there are fewer radicals left for methane to interact with, prolonging its lifetime in the atmosphere and, in turn, its impacts as a major greenhouse gas. The latest study says H_2 has a GWP20 of 33 and a GWP100 of 8 to 12.[1] [2]

2.4.2 H_2 Leaks

The Hydrogen Science Coalition uses information that Michael Barnard and Dr Paul Martin write extensively about the challenges of working with H_2.

- H_2 is the smallest diatomic molecule in the universe.
- Because it is so small, it diffuses through metals.
- H_2 is also the least dense of the elements.
- Effective storage is at:
    - Extreme pressure from 350 to 700 bar (5,000 to 10,000 psi/35,000 to 70,000 kPa), equivalent to 3 to 7 km under the surface of the ocean. That is 1/3rd to 2/3rd the depth of the Mariana Trench.
    - Low, or at cryogenic temperatures. Just above absolute zero at -253°C, in specialised, well-insulated tanks.
- Alternatively, it can be stored at some combination of the two, to have enough of it in one place to be useful.
- No odour, and burns without any colour at all. Even knowing that it's leaking is challenging.

The data on how much leakage is highly variable. The best evidence is that the leakage is about 1% at every touchpoint, or every time it moves from one process to another. [26] For example, for use in H_2 buses, that process involves generation, compression, valves, storage, transport, dispensing into the vehicle and final use, or 8% to 10%. In storage, the loss is about 1% per day, unless the H_2 boiled off is recompressed. There are many instances where real-world leakages have been much higher. A gas station in California had leakage rates of 30% to 35% and after a couple of years of remediation, it managed to get leakage down to 2% to 10%. A liquid hydrogen storage tank at TU Delft in the Netherlands boiled off 5.5% of the hydrogen per day. There are losses with electricity transmission and distribution as well but these losses do not impact emissions and are priced into the supply.

2.4.3 H_2 is An Explosive Mix

H_2 indoors is an explosion hazard. It explodes in a range of 4% to 75% mixtures with air, and at the higher end, that's a lot of energy. About 4,000 buildings go boom today in the USA due to natural gas leaks, after 140 years of trying to make it safer. Over 360 explosions kill 140 people with over 700 injured. Pipeline leaks are a daily occurrence in the US, with over 1.7 per day since 2010. [27] Australia has about 20 LPG and CNG gas accidents per year, with a dozen fatalities. In countries with less stringent safety precautions, older gas bottles and less-maintained distribution systems, the numbers get much bigger. We live with the risks without thinking about them. H_2 is beset with more leaks and with higher energy than with gas.

2.5 Expanded Future H_2 Uses

The good news is that as oil is phased out, emissions from H_2 used in oil production disappear. The majority of hydrogen projects currently being promoted by oil and gas, auto, and industrial turbine companies and countries — which believe hydrogen should be deployed on a wide scale to help the world decarbonise — rely mostly on fossil fuel-derived blue and grey hydrogen. The potential for Australia to become a hydrogen superpower is undeniable, but the path forward is not guaranteed success. Success hinges on the following challenges:

- Overcoming technological challenges.
- Achieving cost competitiveness.

- Navigating fierce competition from alternative energy sources.
- Having a responsible policy that addresses environmental and social concerns.

"The people I respect doing the modelling of this, [they say] you know 90-95% of the economy will be all-electric, with 5%, maybe, hydrogen, on the high side"
Dr Saul Griffin –[28] Electrifying Australia

A special report by Bloomberg Energy Outlook for Australia examines two models for Australia's electricity sector by 2050. One is a net-zero scenario, and the other is an export of hydrogen. This H_2 scenario is large and ambitious, but it is sometimes used to demonstrate how costly the energy transition could be under the net-zero scenario. Their modelling indicates that Australia could produce 28 million tonnes of green hydrogen annually, accounting for 6% of global demand by 2050. Current electricity consumption stands at 210 TWh per year. A hydrogen scenario would require electricity generation to increase sevenfold to 2,000 TWh, representing 170% growth over the net-zero scenario. To achieve this using wind and solar- the most affordable sources of electricity- would require twenty-one times the current solar and wind capacity, compared to double what is needed today to reach 100% zero emissions.

"To supply this demand on a least-cost basis, wind and solar capacity across the country would need to grow to 812 GW by 2050, 21 times more than today." BNEF Energy Outlook Australia May 2023 [29]

To use this much H_2, we need a framework to assess which increase in demand is likely. From an economic perspective, increasing supply without demand leads to failure.

2.5.1 H_2 Ladder Use Cases

Some claim hydrogen is the Swiss Army Knife of renewable energy, so Liebreicht [27] created a hydrogen ladder that frames the narrative about where the existing and future uses of H_2 are.

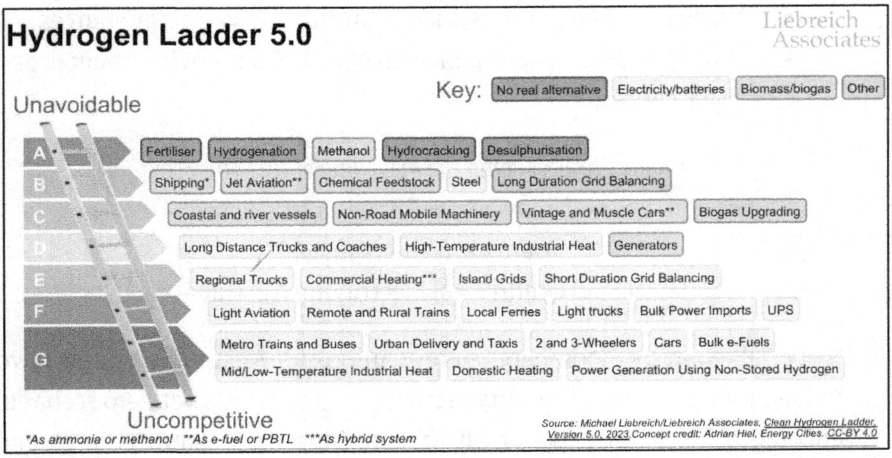

Figure 12 Hydrogen Ladder by Michael Liebreich

Figure 12 of the hydrogen ladder has yellow use cases (D-G) where H_2 is uncompetitive with batteries from transport to heating to grid balancing. Only the top rung of the ladder (A), which includes use cases of fertiliser, hydrogenation, hydrocracking and desulphurisation, has no real alternative. Even the next top rungs (B-C) have other existing options. As the renewable energy disruption continues and costs fall, the use cases where experts said H_2 would have a role will evaporate. More cost-effective solutions win over H_2.

That Australian model [30] discussed in a previous section with expansion from 100mt to billions - for transportation, energy and other uses to replace the fossil fuel industry with a hydrogen-based export industry was not just a tall stretch of imagination, but made the proponents look stupid in hindsight. If they had framed their lofty ambitions with the Hydrogen Ladder, they would have realised that all that demand was nothing but "hopium" – pie in the sky fanciful thinking.

Even worse, they proposed using carbon capture and storage to maintain the fossil fuel industry. They proposed to develop a large carbon capture, utilisation and storage industry to permanently store 80-1000 mt/yr of CO_2 to make clean fuels and negative emissions. As we will see in the CCS chapter, they ignored the reality and economic analysis for wishful thinking. It is difficult to fathom their optimism. Except maybe that they may have been paid by fossil fuel interests?

There are two technical errors many make.

- Failure to understand primary energy. E.g. the value in a car is transport. Most of the primary energy in oil is wasted, with less than 20% actually going into motion. Overall, we waste 65% of fossil fuel energy.
- H_2 is not a swap-out replacement for methane gas, as we will explore in following sections.

2.5.2 H_2 For Heating

Many have said hydrogen can replace fossil methane for heating. Proposals and trials have been done. Dr Paul Martin, an expert in this area, debunks claims that H_2 can replace methane gas. [31] Martin points out the physical and chemical differences between H_2 and methane CH_4. H2 is smaller, with higher leakage rates, higher flame speeds, and burns at higher temperatures than CH_4.

2.5.2.1 Infrastructure implications

- Production: Existing facilities are largely unsuitable for H_2, necessitating new infrastructures.
- Transport: Hydrogen-accelerated fatigue cracking (HAFC) in pipelines poses serious challenges, requiring either blending or complete replacement of pipelines.
- Local Distribution: Leakage and permeation problems are exacerbated in systems using plastic pipes and seals.
- Storage: Due to lower energy density, line pack storage capacity is significantly reduced.
- End-Use Appliances: Required modifications may lead to safety concerns and increased risk of NOx emissions.
- Every time H_2 is moved (production, storage, transport), there is a 1% loss of H_2. Ten per cent loss turns profits into losses.

2.5.2.2 Environmental and Safety Risks

- Transitioning to pure hydrogen introduces serious safety and environmental risks, despite blending offering minor GHG emission reductions.

- Any leakage in the system amplifies safety risks and nullifies savings.

2.5.3 Energy Efficiency

The primary reason why H_2 will not be used for heating is just physics and heat pumps.

- Burning methane – in modern appliances, heating efficiency is up to 95% efficient. Most older gas furnaces are at 60% or less.
- Using electricity directly – 95%
- Using electricity in a heat pump is over 350%. The electricity in the heat pump transfers about 3 to 4 times more energy and the electricity is to pump energy. It is more efficient to burn gas in a power station and use a heat pump than to burn gas directly.
- H_2 needs a lot of electricity to be synthesised from water. The efficiency of electrolysers to produce H_2 is now over 90%, but then you have to pump and transport the H_2 with the associated losses.

Even without physics, economics kill future use. There is a capital cost for the purchase and installation of heat pumps, but they have a lifetime of 10 to 15 years, and even up to 20 years.

2.6 H_2 for Transport

The claimed two major advantages of hydrogen for transport are fast filling (5 minutes), and long range (500 km – 800 km) on par with diesel or petrol. While manufacturers continue to spruik fuel cell electric vehicles (FCEV) there are over ten reasons hydrogen for transportation is dead. Battery electric vehicles (BEVs) have conclusively beaten the claimed advantages of FCEVs, addressing those two advantages, and the only reason we continue to see FCEVs is massive government subsidies and the spruikers having to admit they backed the wrong technology for decarbonisation.

2.6.1 H_2 for Transportation is Dead

Hydrogen for transportation is dead, but like a dead man walking, government subsidies or poor strategic decisions are keeping research and trials alive. They need to be stopped today, and any past investment also stopped.

- Production. Over 99% of all hydrogen produced today is dirty grey hydrogen. It will be at least a decade before green hydrogen is available in quantity, and that date may never come.
- Cost. The cost of green hydrogen would need to be reduced from the current cost of $12/kg to under $2/kg.
- Emissions. The use of grey hydrogen is as dirty in emissions as running an ICE vehicle. There is no contribution from a switch to FCEV to meet climate goals.
- Physics. The efficiency of renewable energy (e.g. solar) for motion in FCEV to BEV is 3 to 4 times less. For every 100kWh of renewable energy, a BEV provides 69 kWh as motion. FCEV only 19kWh of motion.
- Filling stations. In 2022, there were just 685 globally. Even in H_2 countries such as Japan and Korea, filling stations are <0.5% of electric chargers. Public electric chargers are now more available than petrol/gas stations. Private at-home chargers are in the hundreds of millions.
- Scaling. There are about 70,000 FCEVs globally, compared with over 80 million BEV. (2025). That is less than 0.1% of BEV vehicles
- Batteries. Battery costs continue to decrease. A bus company in France [32] discovered cost of FCEV buses would be 6 times the cost of battery buses.
- No Revenue. No ability to generate additional revenue from stored energy. Cars with V2G/V2H or buses can provide energy arbitrage by being charged during the day and discharging at night.
- Limited Servicing. No general skilled repair services for FCEV vehicles.
- Battery charging speed. BEV continues to improve charging speeds. Faster charging (C6), and 1MWh chargers enable 10-minute charging. The latest battery announcement has C20 (3 minute) charging at over 400 Wh/kg energy density.
- Increased range of EVs. The range of BEV has increased. Two years ago, only 3 car models had 350 mile (500km) range. In 2025, hundreds of models are in production.

- Higher density batteries. Higher density means less battery, which leads to lighter batteries. Lighter batteries lead to lighter vehicles, increased range and lower cost of vehicle.

Governments regrettably did not apply subsidies to the process that would use H_2. For example, if a subsidy were on transport, the most effective technology, such as FCEV or BEV, would win. With a subsidy on H_2, more and more subsidies were needed to offset the ever-decreasing battery costs.

A case study for H_2 for buses is in Poland.[33] Polish municipalities embraced hydrogen buses largely because subsidy schemes covered up to 100% of vehicle purchase costs, while electric buses typically received only 60–80% support. Once hydrogen buses entered service, operators discovered the fuel was dramatically more expensive than electricity. By the end of April 2026, Poland had 247 H_2 buses in total, either deployed or contracted by municipalities. 140 buses are already in service, while 107 are contracted but not yet deployed. Despite high hydrogen ambitions, Poland had only nine hydrogen refuelling stations by the end of 2025 against the 32 envisaged by the government's 2021 hydrogen strategy. In comparison, Polamd has more than 12,500 electric charging stations. These buses run on fossil H_2, whereas electric buses run on grids that are 30% renewable and increasing rapidly. Yet even in 2026, the Polish government received on 23 April 2026 a new batch of €7.2 billion EU funds. , of which €500 million is intended to unlock hydrogen investments, particularly to boost private sector growth in renewable and low-carbon hydrogen production. When will this idiocrasy end?

2.6.2 Vehicle Components

Fuel cell EVs are like petrol hybrids. Two power trains. In this case, hydrogen, the fuel, requires a Fuel cell to convert to electricity. [34]. Some companies are banking on hydrogen directly replacing gasoline or diesel, and burning the hydrogen directly, but that has mechanical challenges as we will see.

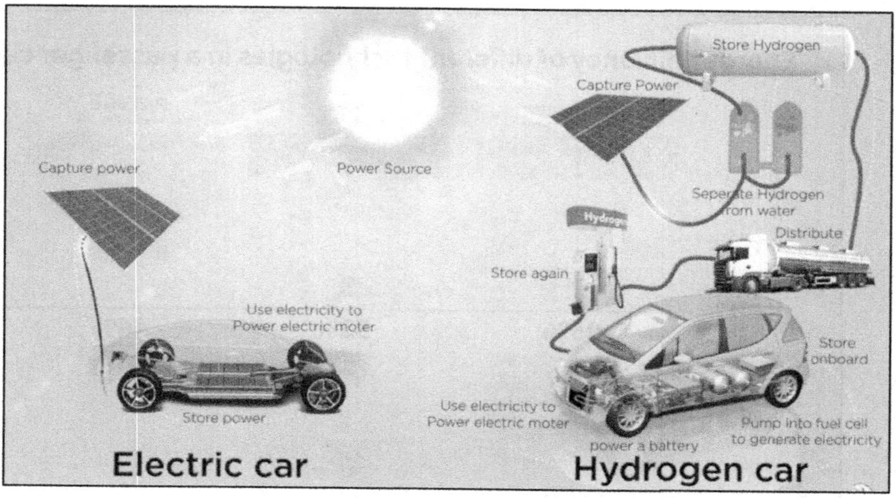

Figure 13 Comparison of an electric car versus a hydrogen car

In Figure 13 we see the comparison between an electric car (EV) and a H_2 car. The EV has a motor and a battery. The H_2 vehicle has an electric motor, a battery, and a fuel cell, or, in the case of a direct-burning H_2 vehicle, an engine. What is also shown but has been ignored is the full supply chain behind the H_2 car. An EV has existing electrical generation infrastructure including a power point in every single 9.1 million homes and business in Australia and for many, all you need is a simple charge plug. The H_2 vehicle system has none of that. No generation. No transport. No filling stations. Just some experimental and pilot ones in some countries. What is really puzzling is that Japanese automakers continue to say that H_2 cars have a place in the world even in 2026, yet there are about 70,000 FCEVs on the road and EVs exceed 70 million. Some say H_2 cars won't have market share, but they continue to state that both light and heavy trucks will run on H_2. The fundamental question is energy efficiency. Efficiency drives the cost of transportation. Cost of manufacturing, cost of operating the vehicle, and the price of fuel.

2.6.3 H_2 Efficiency In Cars

The only thing less efficient than a hydrogen fuel cell (FCEV) is combustion engines using e-fuels made from hydrogen.

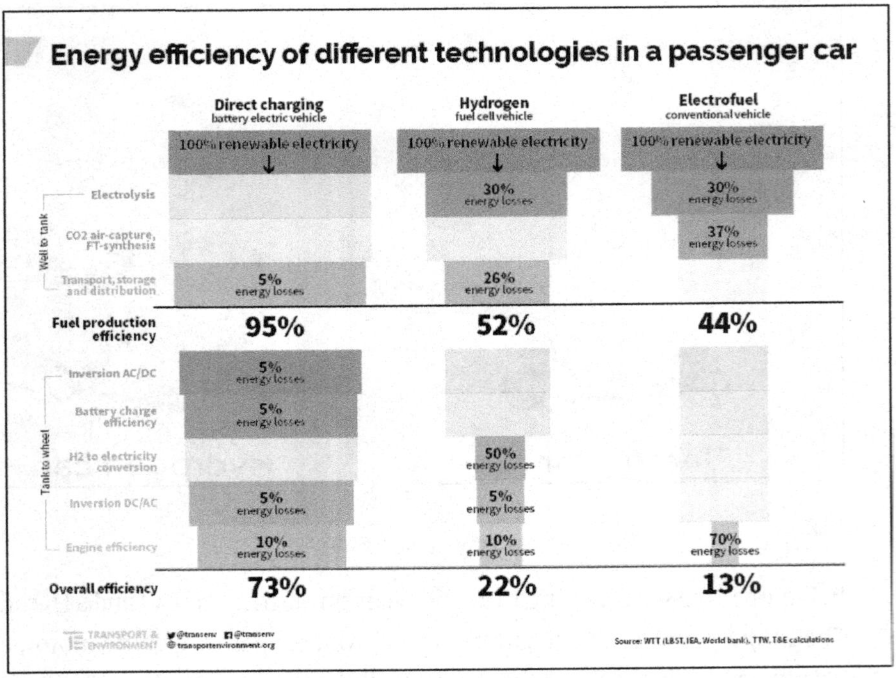

Figure 14 Energy efficiency of different hydrogen technologies versus a battery car

In Figure 14, the authors compared direct charging, a hydrogen fuel cell vehicle and a conventional vehicle using e-fuel. It's hard to call H_2 FCEVs environmentally optimum when they waste 78% of the net energy compared with a BEV at 27% loss. When regenerative braking is included, BEVenergy losses are as low as 8%. The E-Fuel in a conventional vehicle has losses of 87%. In Part 1, we explored the difference between primary energy and useful energy, and how we waste 60% of the fossil fuel as heat. H_2-powered vehicles are inefficient.

2.6.4 H2 Vehicles – Status in 2026

FCEVs are now outsold by Ferraris. There are over 80 million battery EVs on the road now, with over 20.7 million sold in 2025. One in 4 cars sold were EV. The numbers of FCEVs are less than 70,000. By mid 2025, over 50% of new sales in many countries were electric. With an average fleet age of over 10 years, it takes time for the transition of the whole fleet to electric.

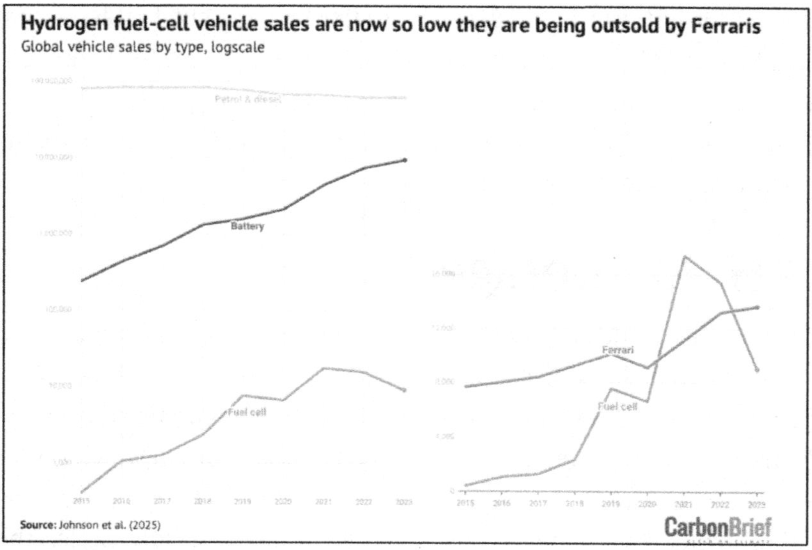

Figure 15 Global sales of conventional, battery electric and fuel cell vehicles from 2015 to 2023.

In Figure 15, we see the global sales of conventional, battery electric and fuel cell vehicles from 2015 to 2023.

- 2023, EV sales = 10.2 million. FCEVs sales = 14,451
- 2024, 17.3 million, FCEV = 14,451
- 2025, 20.9 million, FCEV = 16,011
- The EV fleet has grown from 40 to 80 million
- The inset shows fuel-cell vehicles against the sales of Ferrari models only. [35]. Ferraris, costing over $350,000 outsold FCEVs!

Several manufacturers have announced using H_2 directly in modified internal combustion engines, [1] including Toyota, BMW, Volvo, JCB for off-highway engines and others. H_2 engines burn hotter than petroleum, so engines need hardened valves and valve seats, stronger head gaskets, and a range of other modifications to achieve power outputs comparable to or exceeding those of conventional ICE engines. Without green hydrogen to power the vehicles it appears to be coming to the end of the road, but never get between an automobile manufacturer and a government subsidy.

2.6.5 H_2 for Ammonia for Transport

Ammonia produced from green H_2 has potential for transport use. Green ammonia for fertiliser is discussed in a later section, but Fortescue and others have previously said ammonia has a role as a fuel for hydrogen-powered heavy vehicles and ships. By 2026, Fortescue dropped plans for H_2-powered mining equipment and changed to fully electric. He remains confident for ammonia powered shipping.

2.7 H_2 for Storage

Much has been made of using hydrogen for long-term energy storage. The challenge is when the wind does not blow and the sun does not shine. What storage is needed? Dunkelflaute is a German term that is used in the energy industry to describe a period of multiple consecutive days in which low or minimal energy can be generated by renewable energy sources, such as solar or wind. As chemical grid storage batteries rapidly scale, the conversation is quite different than in 2007 when 100% renewable systems were first proposed, to 2010 with associated enthusiasm for H_2 storage to 2025. In 2010, hydrogen battery storage was considered and promoted as an economically viable alternative to lithium-ion batteries or pumped hydro. H_2 biggest challenge apart from price is associated storage, transport and the exogenous losses.

2.7.1 Long-Term Electricity Storage

As discussed extensively in Part 1, there is a need for long-term electricity storage for electricity grids, such as Dunkleflaut, or for the last 3-5% of grids as they transition from fossil-fuel-based to renewable-energy-based. The relevant questions hotly debated include:

- How much storage is needed? 5%? Less?
- What is the cost to provide this gap?
- Which technology?

Nuclear proponents say nuclear reactors are suitable for reducing storage needs, but electricity generated costs three to five times as much as wind and solar. A model by Dr David Osmond for Australia says 24G W/120 GWh of storage is sufficient for the few days every few years, with the scaling of renewable energy and using existing hydro as dispatchable electricity. [36]. That amount of storage is about 3 hours of demand per day for about 5 days.

A quarter of that is already in operation and construction of the balance is forecast by 2030. Even with that, there is a need for either long-term storage or rapidly deployable electricity generation for a few days a year, as this would provide 98% of the annual electricity demand. More wind and solar decrease this shortage. Compared with Australia, Europe and the USA have more hydro and transmission from other geographies to address these shortages.

A Carbon Brief model says 100% is achievable in multiple cities globally, even in Alaska [37]. They concluded that while achieving 24-hour solar generation is easy in sunny regions, maintaining the same output every single day of the year remains a challenge due to clouds. With the right mix of solar capacity and battery storage, sunny cities can get as close as 97% of the way to full 24/365 solar coverage for just $104/ MWh. No commercial operator will build long-term storage for a few days a year at $100/MWh. They would need a much higher electricity price to achieve reasonable commercial returns. How much do you pay for insurance? What are other options? Lobby groups and storage Operators in Europe have already initiated 9.1 TWh of pure-hydrogen underground storage projects by 2030 and plan to reach 22.1 TWh of capacity by 2040.[38] They claim there is a need for 45 TWh storage by 2030. As an insurance scheme, costs may be less relevant, but nothing like asking for €18-€36 billion. Their intent is to use salt caverns or similar for storage. However, initiation is not completion, and subsidies are drying up. It is worth doing trials, but at this stage the network is still some way from the 97% renewable targets.

2.7.2 H_2 Battery Storage is Unlikely

The EU model to use H_2 as long-term storage is based on multiple assumptions.

The battery would supply electricity for 876 hours. This is based on independent modelling for a 100% renewable grid. Is that hydrogen or other batteries? It is similar to the models for Australia and California but in those grids, lithium-ion batteries are already in use and rapidly expanding.

For this model below, the renewable energy source, H_2 creation, and H_2 burning need to be co-located.

2.7.3 H_2 Battery Costs

This simple model in Figure 12 highlights that hydrogen is probably a financial disaster. The cost of electricity is $253/MWh (high) to $34/MWh (low). Assumptions for operation (pumps, compressors) and the price of H_2 produced at $1 is a fairy tale, and unlikely to meet that price point for decades. It is good that these models are technically feasible, but the economics are against them.

Hydrogen Battery	High-Cost Scenario	Low-Cost Scenario	Notes
Electrolyser Cost	$1,200,000	$300,000	Per MW as per IEA
Production 1 kg H_2	$5.00	$1.00	Cost of H_2 per kg
Energy in 1 kg	33	33	kW energy in 1kg
Conversion to electricity	60%	90%	Efficiency of burning to electricity
kg H_2 for 1MWh	30	30	
Cost per MWh H_2	$253	$34	
Energy differential	$270	$270	Buy RE energy at $30, sell $300 MWh
Margin	$17	$236	After cost of Renewable energy
Hours per year	600	1,000	Usage – model estimates 876 annual
Hrs per 10 years	6,000	10,000	Total offer 10 years
Profit	$104,848	$2,363,300	Inclusive of costs of capital and opex

Figure 16 Hydrogen Battery Cost Model

In 2024, the South Australian Government [39] had promised funding for

$0.5 billion to hydrogen projects. The USA [40] promised over $8 billion. It demonstrates the size of political subsidies. By mid 2025, those subsidies had gone. These subsidies were just props for local politics, and it is a relief to see them gone.

The most cost-effective long-term storage for those few hours may be gas or coal. Countries without gas or oil reserves have moved down the path of using coal to fill the insurance gap. Between the Ukraine conflict and the middle eastern oil disruption, sovereignty has a lot going for it.

2.7.4 Batteries Out-Compete H_2

In this competitive world, battery costs are plummeting. Hydrogen is not being mass-produced, nor will it be for some years. H_2 for storage has all the hallmarks similar to H_2 for transport. Read more about batteries in Optimistic Storm Part 1, The End of Fossil Fuel Era to get some appreciation of battery performance and costs. Batteries continue to improve and are being deployed at an exponential rate. H_2 is still in a negative learning rate phase and is unlikely to be competitive for another decade. Lithium-ion batteries or sodium ion prices are expected to halve in the next decade.

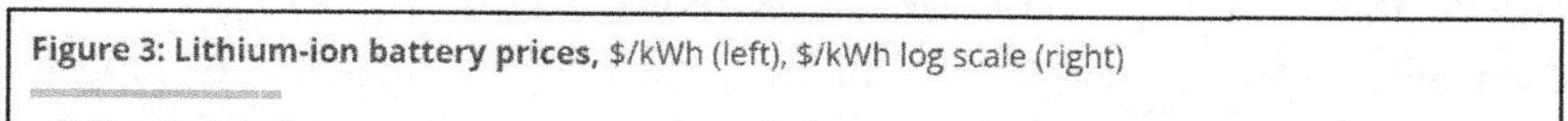

Figure 17 Battery cost decline from 1990 to 2025 and battery market size RMI

In Figure 17, the battery cost learning curve is over 29%. In stark contrast a hydrogen plant needs a large source of renewable energy, extensive civil construction costs with pumps, compressors, refrigeration and tanks. In contrast, a battery has:

- Round-trip efficiencies of 95%.
- Needs only a concrete pad.
- A robust grid connection.
- Flexible. Can be located remotely or on premises.
- Low maintenance costs - no pumps, compressors, pipelines.
- Increasing in lifecycles - now up to 15,000 cycles with zero degradation.

It is pretty clear that batteries have completely outcompeted H_2 for storage. It is not just Lithium or Sodium. An iron battery is in construction to provide 3 GW/300 GWh of storage, or over 100 hours. Thermal batteries in China store heat and then drive the steam turbines for electricity generation and can provide for hours of generation in old coal plants.

2.8 H_2 for Transport of Energy

While high-voltage transmission is generally more efficient for moving electricity long distances, H_2 offers potential advantages in specific scenarios, particularly for energy storage and transportation of energy to remote locations. Hydrogen can be transported via pipelines or ships to areas without existing electricity infrastructure. [41] However, hydrogen has lower round-trip efficiency than high-voltage transmission and faces cost and technical challenges. A review said H_2 for transmission over 1,000 km could be cost-competitive [42], but for shorter distances, HVDC would be cheaper. Others say H_2 for transmission is as expensive as HVDC. HVDC and HVAC already integrate into existing networks. A whole new management system will be needed for H_2. Exergy (turning H_2 back into electrical energy) is inefficient. With large subsidies, Europe built the first segment of the pipeline, extending nearly 400 kilometers from Lubmin on the Baltic Sea to Bobbau in Saxony-Anhalt. They have plans for over 9,000km. Eventually, about 40% of these pipelines will be brand new. The other 60% will involve converting existing natural gas conduits. Of the first 525 kilometres, 507 kilometres come from existing gas lines. There is just one problem – there is minimal green hydrogen. If you view the global prediction over the past 30 years, as in Figure 7, versus the reality, one could have complete derision. Just don't invest in H_2 get rich schemes.

2.9 H_2 For Fertiliser

Fertilisers are responsible for approximately 1.8% of global carbon dioxide emissions. Ammonia, the primary fertiliser, is produced by the Haber-Bosch process. Hydrogen reacts with atmospheric nitrogen over a metal catalyst at high pressure and temperature. Green fertiliser uses the same Haber-Bosch process, but requires both green H_2 and green electricity.[43] China is the largest producer of ammonia, accounting for 30% of production (and 45% of CO_2 emissions), with the US, EU, India, Russia and the Middle East accounting for a further 8-10% each. Ammonia is traded around the world, with global exports equating to about 10% of total production. [44]. Manufacturing emissions are not the only source. The application of fertilisers in agriculture also leads to nitrous oxide (N_2O) emissions, a potent greenhouse gas. Agriculture contributes significantly to total anthropogenic N_2O emissions. CSIRO found that, globally, fertilisers and the management of livestock manure are leading to increased N_2O emissions and rapid accumulation in the atmosphere. This is more than a climate problem. N_2O also depletes the ozone layer, which protects humans from harmful solar radiation. And nitrogen runoff from fields pollutes waterways, increasing harmful algal blooms and creating oxygen-depleted dead zones. [45]

2.9.1 Alternatives to Ammonia Fertilisers

Could nitrogen fertiliser be replaced by AI and bacterial techniques? Is this feasible? Some say the emissions in the entire process are not 1.8% but closer to 7%, from production to the crop. The USA alone applies 6 million tonnes of nitrogen to corn. Is it possible to engineer a bacterium or process that can live inside corn or wheat and convert nitrogen from the air into plant-available nitrogen nutrients? Can scientists replicate the natural conversion process seen in legumes?

It turns out that companies are trying to do just that. Ginkgo Bioworks [46] has a project with Bayer (from 2022) that uses their AI designer process to create the nitrogen nutrient within the plant. Hellriegel and Wilfarth discovered the nitrogen-fixing process for legumes in 1886. Biochemistry has sorted out the pathways. The task given to Ginkgo was to make a bacterium, or to introduce the genetic and chemical pathways into an organism that can live in the plant and supply nitrogen for growth. Their process was also investigating which of the multiple pathways to seed the process. Was the bacterium or organism a seed coating, an application delivered during crop growth, or some other delivery service? Pivot Bio uses microbial products that enable crops to draw nitrogen from the air, giving farmers something they've wanted for a long time: a more reliable and efficient form of fertiliser. [47] Their solution, which produces less than 1% of the emissions from synthetic fertilisers and requires 1,000 times less water, is already being used across 2 million ha of land in Brazil. [48]

2.9.2 Economic Impact Pivot Bio is Profound

The estimated global size of the nitrogenous fertiliser market is $166 billion in 2024. The growth is expected to reach $232 billion by 2030, with a CAGR of 5.7% during the forecast period (2024-2030). [44] Of the 5 billion ha of agricultural land, 1.8 billion hectares are croplands. This is 12 % of the global terrestrial area.

Could a new process disrupt not only the existing conventional fossil-fuel-powered fertiliser process but also the new green process that relies on green hydrogen? Michael Barnard [49] estimates that 30 million tonnes of hydrogen are used in existing fertilisers. Either emissions are reduced in the fertiliser process by green H_2, or emissions are reduced during the crop-growing cycle. Barnard provides some numbers around the emissions avoided.

- Growers can reduce fertiliser by 40 kg per ha or about 20-25% of the fertiliser they would normally use when they apply Pivot Bio.
- There is about 9 kg CO_2e per kg of fertiliser.
- I.e. 360 kg of avoided CO_2e per ha.
- Pivot Bio had over 0.4 million ha under management in 2022.
- That means over 180,000 tons of CO_2e was avoided for the 0.4

millon ha.

- About 1.8 billion ha of cropland in the world.
- Assuming 80% penetration of Pivot Bio or equivalent product, that would be over 660 megatons of CO_2e avoided per year and at full production closer to 1 Gigatonne.
- This technology could deal with a 1/3rd of fertiliser emissions and slash ferti

2.9.3 Reduction in Fertiliser Demand From Ethanol

Pivot Bio is just one company with an approach. It will reduce usage, but the biggest change to the amount of fertiliser use will not be within the existing agriculture system but may be disrupted from outside the industry.

- Over 40% of corn production in the USA is for growing ethanol. [50] Although 40.5% of corn grain was channelled to ethanol processing in 2011, only 25% of US corn acreage was attributable to ethanol when accounting for feed co-product utilisation. By 2026, land area attributed to corn ethanol production is reduced to ~15% depending on the corn grain yield scenarios.
- As vehicles change to electric over the next two decades, production of ethanol for land transport will collapse to zero. What will that ethanol be used for?

2.9.4 Reduction in Demand from Precision Fermentation

Precision fermentation will reduce the cropland needed for other grains and animal feed. Various estimates that the area of cropland will shrink by 80% within 20 years as precision fermentation disrupts the feed to animals to humans. See Part 2 Optimistic Storm.

The amount of hydrogen needed to make fertiliser will tumble. Ammonia used for explosives will also fall as coal mining falls due to the rise of solar and wind electricity generation. The area for mining for minerals will fall by 80% as fossil fuel exits electricity and steel manufacturing.

2.10 Plans vs Reality

The media is full of hydrogen plans.[51] The reality is that the anticipated demand and supply are thousands of times higher than current initiatives. The whole process is a complete, fanciful dream, and regrettably, is used by anti-renewable critics to say that renewable energy is a con. Based on the evidence presented here and other places, hydrogen was badly over-hyped. With the current understanding, we are now coming back to a realistic understanding of future appropriate use cases.

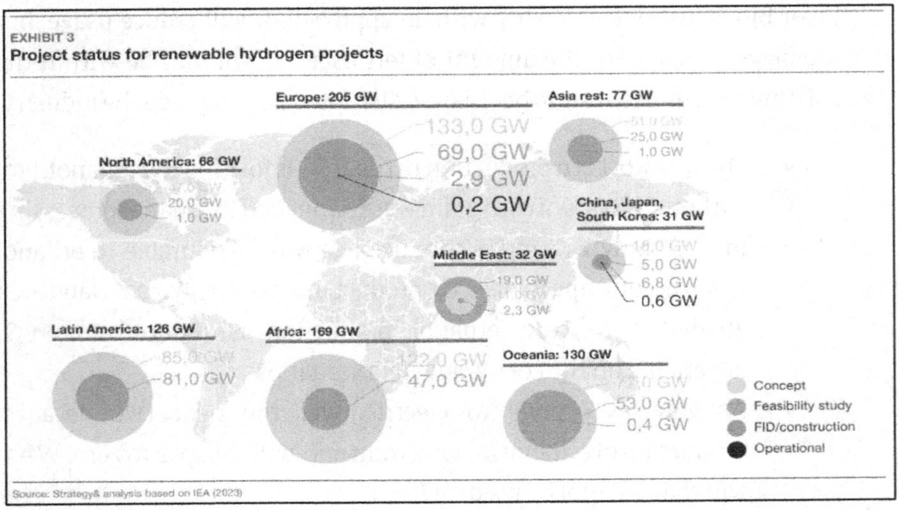

Figure 18 Renewable hydrogen projects planned vs actual plants

In Figure 18 it shows Europe needs 133 GW by 2030. The actual is probably 0.2GW or less than 0.15% of the forecast. That is 677 times a misjudgment. One has to ask. If the plan is a complete fantasy, are they really serious about transitioning to clean energy and zero emissions?

2.11 Hydrogen Cult Madness

In 2026, H_2 is having a well-overdue wake-up call. Michael Barnard says six groups have an H_2 focus. [52]

- Oil companies want to transform oil reserves to H_2.
- Banks and institutions that fund oil companies.
- Governments that have little science (STEM) and lots of lobbyists

and primary and secondary impacts on GDP.

- People and companies who have invested in H_2 and already lost money.
- Companies that are dead, but still moving around to cash in on Government handouts. The IRA during Biden's presidency in the USA added subsidies on top of subsidies for a mountain of subsidies.
- Credulous fans, who are almost entirely devoid of STEM understanding.

Electrolysis-produced hydrogen isn't a straight route to decarbonisation; it's more like driving in circles when a shorter, direct path via electrification is available.
Michael Barnard

2.12 Conclusion: Hydrogen

Hydrogen has captured the minds and pockets of politicians. The promise to be all things to all clean-emissions has turned out to mean that the politicians backed the wrong horse. The only country continuing with transportation is South Korea, but the USA, EU and Australia are continuing to flood the market with hydrogen subsidies for other uses. Companies, e.g. Fortescue, were leveraging their politics and global reach to set up green hydrogen hubs. Data says has to be very close to the end user. Those users are not there. Many announced plans fell in a heap in mid-2025 and even more in 2026. The production, transport, and storage of hydrogen pose technical and economic barriers. Promotion is for political purposes. There's now healthy scepticism about the role of hydrogen in the energy transition, and replacing grey hydrogen with greener alternatives will be costly and often impracticable. In summary.

- Half the current use is for oil and gas refining. Replacing those fossil fuels with electricity from renewable sources halves future demand for H_2.
- There are no new uses identified.
- Hydrogen is not a universal energy solution.
- Domestic heating is now dominated by heat pumps powered by

renewable electricity, which are 3 to 5 times more efficient than gas. And gas has higher efficiency than H_2.

- Fossil Fuel Greenwashing. The fossil fuel industry is using H_2 as a bait-and-switch to protect stranded gas assets and business models.
- Cost of production. Green hydrogen depends on electricity costs. It was thought there would be surplus free electricity during peak solar production. The rise in cheap batteries means they can store the excess midday solar energy and move it to the after-hours supply. That whole concept of free midday electricity has evaporated. In the long term, electricity prices will make H_2 too expensive.
- Electrolysers. Costs have fallen but electrolysers are only one component.
- Associated Manufacturing Costs. Costs of the mechanical compression, and transport are unlikely to reduce much.
- H_2 is not suited to land transport. H_2 is inefficient and costly for vehicle transport. We will continue to see hydrogen-powered vehicles for a while until the operators run out of money and sense.
- Use in steel making has been a primary use case, but the only way it will be economic is with a carbon price. The next chapter will show other technologies may be a competitive threat. The steel from stranded assets such as oil pipelines, refineries, ships, and infrastructure will be fed into existing steel manufacturing, reducing the demand for iron ore for years to come.
- Use of H_2 as a synthetic fuel is not competitive with biofuel or batteries.
- There is no economically competitive options for some hard-to-abate emission sectors such as long-haul aviation.
- The current subsidies on hydrogen are badly misplaced. Most countries subsidised H_2 directly instead of the end process.
- Hydrogen is a distraction from decarbonisation and is going the way of other "promising" technologies.

All we are going to see now is a slow death, as fossil fuel companies continue to push H_2 usage over the next decade. Betamax lost to VHS and it was all over in 5 years. We may have to wait for a decade for hydrogen.

It will take until 2030 for the hype to die down, because that's how long it takes to deprogram a cult.

Dr Paul Martin - Hydrogen Science Coalition

Conclusions of the Hydrogen Folly:

- Start with understanding of H_2 physics and economics.
- The Hydrogen Ladder helps reframe its use.
- Governments and investors ignored science and economics.
- Fossil fuel companies are looking for alternatives to burn methane
- H2 is a poor no substitute for methane.
- It will 20 years hydrogen to fall to low levels of use. It will only be used for niche uses.

3 Green Steel

Green Steel:

- Steel is a foundation of our modern society.
- Emissions from fossil fuel used in steel production is about 7% to 9% of global emissions.
- We will explore the use of H_2 in green steel as a specific example of reality and hype.
- Review other low emission options for steel production .

Steel is an integral part of our developed civilisation. Americans have 16 tonnes of steel per person, China 8, and Africa 1. Aside from the household car (1 tonne) to hospitals, railway, schools, steel per capita may be a better indication of wealth than GDP. Steel is currently produced worldwide at a rate of 2 billion tonnes per year. For every tonne of steel produced, the steel industry emits 1.9 tonnes of CO_2, contributing roughly 8% of the global anthropogenic carbon emissions. A large part of these CO_2 emissions (over 70%) occurs during the reduction of iron ore to pure iron or pig iron, which is the feedstock for steelmaking. Currently, iron ore reduction is primarily performed in blast furnaces, using a carbon-based reduction agent, like carbon monoxide (CO) released from coke. This reduction agent reacts with the iron ore, forming CO_2 as the by-product:

$$Fe_2O_3 + 3CO \rightarrow 2Fe + 3CO_2$$

Green H_2 is one proposed solution to decarbonise the steel-making process.

But like all solutions, removing emissions from steelmaking is complex. There are already solutions, and the proposed large quantities of H_2 needed for the future are questionable. Certainly the Global South will need more steel in its development from poverty. There may be alternative solutions, but the scale of decarbonisation is vast – over 1,400 blast furnaces are currently in operation globally. Global Energy Monitor details the 1,408 blast furnaces across 477 plants in 55 countries. Some are no longer operational. [53] Most new planned steelmaking is low-emission electric arc furnaces.

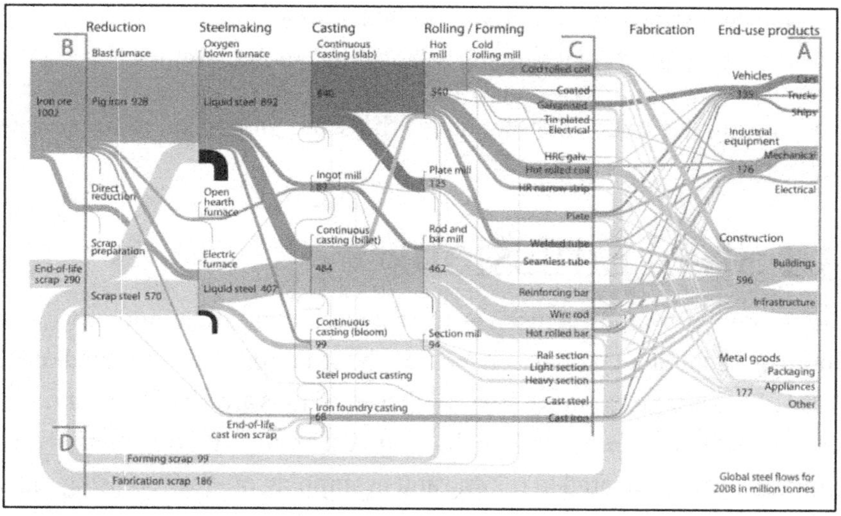

Figure 19 Global steel flows for 2008 in millions of tonnes by Julian Allwood 2016

Figure 19, though complex, details how complicated steel production actually is. The thickness of the lines is relative to the quantities. Across the top of the graphic are the various processes we will look at in more detail.

- Reduction – the actual iron oxide to iron process.
- Steelmaking – turning iron into steel.
- Casting – putting that steel into bars, or strands.
- Rolling and forming – pulling those bars, strands, or ignots into steel products.
- Fabrication – turning those products into end-use products you buy, from construction to trucks to packaging and domestic appliances.

Globally, crude steel production totalled 1,951 million tonnes in 2021. This production is spread all over the world, including India, Japan, and the USA, with the vast majority (1,033 million tonnes) concentrated in China. The average life of steel is over 38 years. Iron ore accounts for 2/3rd of steel (1,000 million tonnes), and the balance is end-of-life scrap (570 million tonnes). Based on the figure above:

- More than half of the steel is used in construction – 596 million tonnes.
- Vehicles – 139 million tonnes.
- Industrial equipment – 176 million tonnes.
- Metal goods make up the balance – 177 million tonnes.

Forecasting out to 2050 shows about the same proportion of iron ore vs scrap. [54]

3.1 Steel Emissions

Some processes lend themselves to easy decarbonisation, and others are more difficult. Steel production creates about 1.8 tonnes of carbon dioxide (CO_2) emissions per tonne of steel. Global emissions are 3.2 billion tonnes of CO_2 equivalent to 403 million households or nearly 700 million cars. Emissions can be abated in mining and in shipping ore to processing mills, but the largest source of emissions is burning coal to remove the oxide from iron ore to produce steel.

We will review the reduction methods in the next sections.

- Hydrogen-based Direct Reduced Iron -DRI.
- Use electric Arc Furnaces – EAF.
- Molten Oxide Electrolysis – MOE.
- Biomass oxide reduction (swaps coal for biochar).
- Microwave hydrogen plasma.
- Carbon Capture (CCS is a scam as we see later).

Aside from the reduction in emissions in the steel-making process (and H_2 has been the most talked-about), there are other emission-reduction actions with solutions today.

- Zero emissions mining. Companies such as Fortescue are working to have a fully zero-emissions by 2030 by replacing all the diesel used in mining. Scope 1 and 2.
- The amount of steel in stranded assets that will come free as era 4 energy rolls on is enormous. Pipelines, refineries, mines, fossil-fuel power stations, and ships (45% of global tonnage) would free up enough steel to meet a decade of steel demand. These use electric arc furnaces and could be zero emissions.

At this stage, H_2 increases the cost of making steel. A carbon price is the only way to reduce the imbalance between dirty steel and green steel. There are various pilot plants worldwide, but demand for green steel remains low. Most steel producers are waiting for better technology before investing in change.

3.2 Steel Reduction Using Coal

Coal is used for 2 purposes. (a) Processed coal (also known as coke) is used in the main body of the furnace, acts as the physical support structure for materials entering and leaving the furnace, among other functions. (b) It's also partially burnt at the bottom of the furnace to both produce heat and make carbon monoxide, which strips oxygen from iron ore, leaving metallic iron. (11). Hydrogen can replace the CO, but it's not 1:1. More heat is needed, and the solid structure provided by coal also needs an alternative.

Steel Production Method	Materials Utilised	Steel Making Capacity (mtpa)	CO_2 Emissions (2019)
EAF - electric arc furnace	Scrap	701	0.5 Gt
BF-BOF blast furnace, basic oxygen furnace	Scrap, iron ore, coke	1,483 (67%)	3.1 Gt (86%)
Totals		2,183	3.6 Gt

Figure 20 Steel-making processes, fuel and CO_2 emissions

Figure 20 shows the two current methods. Both the traditional electric arc furnace (EAF) and blast furnace basic oxygen furnace (BF-BOF) production use fossil fuels. The blast furnace process was 67% of steel but 86% of emissions. A hybrid green steel method uses hydrogen to reduce iron pellets to sponge iron, then to steel. The process requires high temperatures but below the melting point of iron (800-1,200°C), saving energy costs.

3.2.1 Transformation

Pig iron is alloyed with other minerals such as nickel, chromium, manganese, or molybdenum. This creates a crude steel material with the physical properties required in the final application. Most often, fossil fuels provide the energy needed, but direct electricity could be used.

3.2.2 Forming

Steel is formed through a series of hot and cold processing steps into the final shape that the customer will use. As with transformation, this energy-intensive step could be powered by green electricity.

3.3 Direct Reduced Iron (DRI)

Decarbonising using hydrogen instead of coal adds substantial capital costs and operating costs to the iron ore reduction process. Not only does one have to produce and transport enough H_2 for the DRI process, but one also has to invest in new production facilities, because coal/coke used in blast furnaces provides the infrastructure inside the blast furnace.

- Process. Reduce iron ore with hydrogen instead of coke.
- Technology. Requires new DRI reactors and a H_2 supply.
- A hybrid system, at 800°C, would substantially reduce emissions.
- Projects. HYBRIS (Sweden), HYFOR (Germany), H2 Green Steel (Sweden), Primetals Technologies (Germany).
- Advantages. Potentially near-zero carbon footprint with green hydrogen, high flexibility in fuel sources.
- Challenges. An enormous quantity of expensive H_2 and associated infrastructure development for hydrogen transport and storage.

Figure 21 shows the enormous amount of renewable energy needed to switch the annual steel production to a green H_2 DRI process.

Steel Production Source	Annual Steel Production	Green Hydrogen Required	Electrolyser Capacity Required (GW)	Total Renewables Capacity Required (GW)
Base reference	1 Mt	50 kT	0.56	0.7
U.S.	85.8 Mt	4.3 Mt	48	60
Europe	103 Mt	5.2 Mt	58	72
China	1032.8 Mt	51.6 Mt	581	726
Global	1951 Mt	97.6 Mt	1,097	1,371

Figure 21 Steel production source annual steel production and total renewable energy capacity required

Electricity from 1371 GW is about 12,000 TWh, which is roughly the total renewable electricity produced today. In context, the global energy use is about 166,000 TWh per year. Change is technically feasible. For example, in 2025, the total new investment in renewable electricity was 814 GW. The challenge is economics. DRI using H_2 is about ten times the cost of the current processes because H_2 is about $10/kg, and the cost needs to fall to about $1-2/kg.

Green hydrogen is too expensive to use in our EU steel mills, even though we've secured billions in subsidies.
Geert van Poelvoorde, ArcelorMittal

3.4 Electric Arc Furnaces (EAFs)

There is no need to wait for 100% renewable hydrogen production to decarbonise steelmaking, since the current global average grid power is sufficiently low-carbon to break even with conventional electric arc furnaces.

- Process. Use recycled steel scrap and melt in electric furnaces powered by renewable energy.

- Technology. EAFs, scrap processing facilities, and renewable energy sources are all existing technologies, so the only challenge is the economics and whether the consumer will pay for green electricity. Older EAFs would need upgrading.
- Existing projects include Big River Steel (Arkansas, USA), Nucor (USA) and ArcelorMittal (worldwide).
- Advantages. Leverage existing infrastructure. Lower investment compared to new greenfield plants.
- Challenges. Limited to using scrap and cannot produce steel from scratch, and relies heavily on the availability of renewable energy.
- Renewable energy availability is increasing, and within a decade, most major electricity grids will be predominantly renewable-sourced.

3.5 Molten Oxide Electrolysis

Boston Metals has developed a molten oxide electrolysis (MOE) process. This direct approach eliminates several steps in the steelmaking process and does not require coke production, iron ore sintering and pelletising, blast furnace reduction or basic oxygen furnace refinement and [55] the electrolysis process eliminates coal. Just electricity is used for the conversion of all iron ore grades to high-quality liquid metal. The process takes place in an MOE cell, where an inert anode is immersed in the electrolyte containing the iron ore. High electrical currents increase cell temperature to 1600°C. At that temperature, electrons split the bonds in the iron oxide in the ore, producing pure liquid metal. No carbon dioxide or other harmful byproducts are generated. There is no waste. The only product is oxygen from the iron ore (Fe_nO_m), and the process needs no water, no hazardous chemicals, and no precious-metal catalysts.

Sounds too good to be true – but like much innovation, it has taken a decade to scale up from lab scale to pre-production. Boston Metals needs to scale more rapidly. At what cost? Most recent reports suggest that while they are still progressing steel, other metals such as niobium, chromium, and manganese are much more lucrative. [56]

3.6 Biomass Reduction

Various industries have developed biomass-based reduction systems. It too faces challenges and requires new DRI reactors, and while able to reduce emissions somewhat has other challenges.

- Process. Use biochar or biogas instead of fossil fuels to reduce iron ore.
- Technology. Biomass gasification plants, DRI reactors, and furnaces for melting DRI.
- Projects. SSAB (Sweden), voestalpine (Austria), BlueScope steel (Australia).
- Advantages. Renewable and potentially carbon-negative source of reduction agent.
- Challenges. Biomass availability and sustainability concerns lead to higher emissions than green hydrogen.

3.7 Microwave Hydrogen Plasma

Another innovative zero-emission technology is microwave hydrogen plasma. Researchers claim that rapid reduction of iron oxide to iron is possible within seconds, and that the plasma method is faster than thermal reduction due to the energetic plasma species. [57]

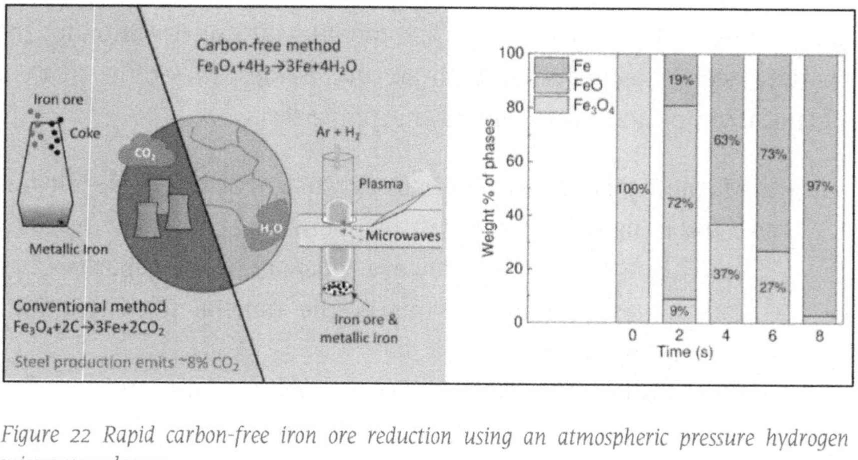

Figure 22 Rapid carbon-free iron ore reduction using an atmospheric pressure hydrogen microwave plasma.

Figure 22 shows, on the left, a conventional blast furnace, which emits 8%

CO_2. On the right, the iron ore, argon and H2 provide metallic iron. At 8 seconds, 97% of the iron ore has been reduced to iron. This research provides a potential novel way – but it has to be scaled from lab to industrial scale, and that will take years. This novel process, like MOS, is promising, but plasma is well behind MOS, and MOS took a decade. Moving a whole industry is difficult, and in a decade who knows what new technologies will surface.

3.8 Carbon Capture and Storage

Some consider that carbon capture and storage (CCS) may be suited to steel production, but as we will see in Chapter 5, CCS is a failed technology and unlikely to be a viable solution to capture CO_2 emissions from traditional blast furnaces and store them underground. The technology is complex, the CO_2 stream is not concentrated, plants are old and retrofitting CCS is unlikely to be an economically viable solution. You cannot make a silk purse out of a sow's ear is the expression that comes to mind.

3.9 Different Iron Ores

A complication with any whole-of-industry change is that there are complications at every step. A critical and misunderstood aspect of iron ore itself is the complexity. The type and concentration determine subsequent processing steps. Magnetite and hematite are the two most important iron oxide minerals that make up iron ore, the primary source of iron for the world's steel industry. While both are iron oxides, they differ in their chemical composition, iron content, and properties.

- Magnetite (Fe_3O_4) is a black, strongly magnetic iron oxide mineral. It has a higher iron content than hematite, potentially reaching 72.4% iron when pure. However, magnetite ore deposits often contain lower concentrations of the mineral itself, requiring concentration before use.
- Hematite (Fe_2O_3) is a reddish-brown iron oxide mineral that is also a primary iron ore and is generally the one most often used in blast furnaces. It generally contains 60-70% iron when pure. Hematite results in lower iron content and higher impurities. Hematite is

crushed, separated out and smelted.

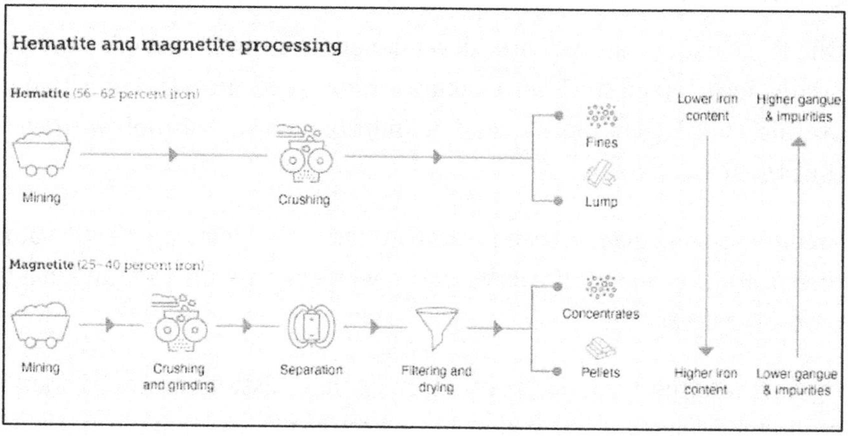

Figure 23 Comparison between hematite and magnetite processing. Australian Industry Energy Transitions Initiative

Figure 23 shows the processing steps for hematite and magnetite. Magnetite is crushed, separated from the rock, filtered and then processed. Magnetite has higher iron content and less impurities. Over 96% of Australia's exports are hematite, as it has a high iron content and is abundant. [58] On the other hand, Australia has extensive magnetite reserves, but the additional pre-processing costs have made them more expensive. Fortescue's recently developed Iron Bridge project has started production. A MOS process would further open up that source of iron ore.

3.10 Mining & Transport of Ore

Scope 1 and Scope 2 emissions already have technology solutions to reduce emissions from steel production. Companies such as Fortescue [59] are electrifying all heavy equipment used in mining and gas used in preprocessing. They plan to reduce diesel consumption by over 500 million litres each year by 2030, converting their diesel excavators, dump trucks and rail systems to zero-emission electric vehicles. Note that emissions from scope 1 and 2 are minor compared with scope 3, which is up to 97% of the total steel emissions.

3.11 Conclusion Green Steel

One of the major reasons for the development of green H_2 was enabling the production of green steel. Developing technology to direct the reduction of iron ore using H_2 has been successful. The economics are horrible with a ten times increase in costs.

Green H_2 is reducing less than 1% iron ore currently. Unless there is a major technical and economic breakthrough, it won't be anytime soon and that is probably 2 decades.

In the meantime, the industry can clean up the emissions in scope 1 and 2 – mining and transport, and major changes can be made within the processes outside of the reduction process to halve the emissions.

Conclusion of Green Steel:

- Eliminating emissions from steel production is an expensive and slow process.
- Reduction of iron ore with H_2 is in production but uneconomical.
- Other novel technologies are promising but a decade away.
- We need to continue to invest in research and early production methods.
- Investment is best deployed in electrifying other processes in steel such as mining along with using renewable energy used in steel making.

4 Nuclear

Exploring Nuclear:

- Nuclear power was developed as an alternative to killing people.
- The rise of nuclear in the 1950s and 1960s has been matched by the equally dramatic retirement of those plants.
- There is no "nuclear renaissance", and the increase in 2024-2026 is less than 5 GW whilst renewable energy has surged by over 1,000 GW.
- We will assess radiation, waste and accidents.
- Taxpayers subsidies are better spent to increase renewable energy faster and for longer.
- Nuclear - remain as a niche electricity source.
- Within 2 decades most nuclear will be replaced with wind solar and batteries.

We are going to take the view in this book that conventional nuclear is in its dying days. All that current nuclear activities do is prolong decisions until it becomes obvious that the nuclear age is over. Others may disagree. This chapter lays out the technical and economic perspectives. Nuclear is mostly about politics. From an emissions perspective, we do not have 18 years for nuclear to decarbonise the planet. An old technology, portrayed as new, has made little progress and can be explained by the principles of innovation in Chapter 3 of Part 1 of Optimistic Storm. For example, fusion, though still 20 years away, may solve the energy demands of the future. The joke in the industry is that for 50 years, fusion has been $20 billion and 20 years away.

When nuclear power was first invented, the claim was that nuclear power would be so cheap that you would not have to meter it. Fifty years on, that's no longer true, and taxpayer subsidies primarily fund the industry. New nuclear plants are an inefficient use of scarce investment capital, and renewable energy offers better returns. Emissions will reduce with existing clean, safe and low-cost renewable energy. Where is support for new nuclear-based power coming from?

- Some by the fossil fuel industry want to slow down renewable energy uptake to avoid the loss of coal and gas.
- Some want to hold onto the past.
- Some genuinely believe nuclear power has a role.
- Some want the knowledge for development for military use. Most countries that have adopted atomic power for power production have inevitably used their nuclear facilities to build out their military expertise.
- Some lack technical and economic knowledge that wind, solar and batteries are the cheapest and fastest sources of clean energy.

We will try to keep concepts simple in this chapter and not delve too much into physics although sometimes we will need to delve into nuclear physics to see how advocates are distorting the reality. We generally will use the word uranium but it is complicated. uranium-235 (or ^{235}U) is an isotope of uranium and makes up about 0.72% of natural uranium. Unlike the predominant isotope uranium-238, it is fissile, i.e., it can sustain a nuclear chain reaction. It is the only fissile isotope that exists in nature as a primordial nuclide and was discovered in 1935. It has a half-life of about 704 million years. When it decays, it can create a whole raft of other radioactive isotopes and elements, including astatine, bismuth, francium, polonium, protactinium, radium, radon, thallium, thorium, and eventually lead. [60]. A nuclear reactor harnesses the energy in a controlled manner – a nuclear bomb does not. The resulting fissile elements also differ depending on whether they are controlled in a power plant or detonated as a bomb.

4.1 Uranium to Electricity

Most have little understanding of how we start with uranium in the ground and wind up with electricity production. When we dig even a little into the technology needed for this pathway, we see why this chapter comes to the conclusion that new nuclear power will play only a minor role in the future. Each of the twelve steps faces complications and challenges.

- Prospect and locate a uranium source. There are about 97 mines, producing 50,000 tonnes, with Kazakhstan, Canada and Namibia producing 70% of global supply. Some is conventional ore, others are in-situ leaching. [61]
- Mine or extract that ore-grade uranium. I.e. Inert U-238 plus the 1% U-235.
- Concentrate ore/extract.
- Ship to an enrichment refinery.
- Enrich.
- Make up fuel rods.
- Construct a nuclear reactor.
- Create a radioactive reaction.
- Operate a nuclear reactor.
- Build and operate a steam generator using the heat from the nuclear reaction.
- Store the waste from the leftover fuel.
- Dismantle the nuclear power plant. Store the highly reactive components.

What's my expertise as an author? While I did not spend my career in nuclear, I did formally study nuclear engineering and worked in a research nuclear reactor. My PhD in Weed Science in the early 1980s used ^{14}C to investigate the mode of action of herbicides, employing a 4MW reactor for sample bombardment and analysis. To do that research, the NRC required a course in nuclear engineering. This was well before Homer Simpson's nuclear safety parodies.

4.2 Nuclear Rise and Fall

Russia's first nuclear power plant, and the first in the world to supply electricity to the grid in 1954 was the 5 MW Obninsk reactor, which supplied electricity until 2002. The global 440 nuclear plants, many constructed in the 60's, will require decommissioning by 2050. Construction of the only new plant in Europe, Hinkley C, is delayed by 10 years and $28 billion. Nuclear plants using low-enriched Uranium require refuelling about every 10 years, costing hundreds of millions, and the downtime takes up to 18 months. Back in the 1950s, people assumed nuclear power could supply 20% of the world's energy, and grow to over 50%, but that did not materialise for the following reasons.

- Ongoing accidents. Two severe accidents have caused widespread injuries and deaths.
- Multiple countries converted their fuel for reactors into military use and created atomic bombs.
- Waste. Countries have failed to sort out waste.
- Social license lost. Widespread pollution from military testing in the Pacific, the USA, and Australia eroded social acceptance.
- Only governments will fund new plants.
- Construction of new plants is too slow to use for decarbonisation.
- Existing nuclear is now the most expensive form of electricity.
- Wind, solar and batteries are alternative solutions and provide much cheaper electricity.

Like many innovations, nuclear suffers from problems within and competition from outside. Nuclear technology in the 1950s could have expanded like other technologies, but regulation following a series of accidents killed further investment and growth. [62] We will look at the current excitement about nuclear in 2026, but these issues have not magically disappeared.

4.3 Top 15 Myths of Nuclear

Any time you read an article promoting nuclear power, one or more of these myths will be included. The investment in AI and its apparent insatiable demand for more electricity is particularly bad at promulgating myths. A later section will assess the economics of AI that drives focus back onto nuclear as an electricity source to meet that demand.

- Base-load power stations are necessary. Or a variant of the myth that base-load stations back up variable renewable energy.
- There is a renaissance in nuclear energy, and countries around the world are now investing in new nuclear.
- Renewable energy is not ready to replace fossil fuels.
- Nuclear weapons are independent of civil nuclear energy.
- Nuclear is low risk and not dangerous.
- The death toll from accidents was trivial.
- The problem of permanently storing nuclear waste has been solved.
- Nuclear emits no or minimal greenhouse gas emissions.
- Nuclear energy is a suitable partner for renewable energy in the grid.
- Nuclear reactors can be operated flexibly to follow changes in demand or load.
- Firmed renewable energy is more expensive than nuclear.
- Renewable energy is very diffuse and hence requires huge land areas, while nuclear does not.
- Energy payback periods (in energy units, not money) of renewable energy technologies are comparable with their lifetimes.
- Danish / name your country/electricity prices are among the highest in Europe, because of the large contribution from wind energy. Hence, they should use nuclear power to reduce costs.
- Computer simulation models of the operation of electricity grids with 80-100% renewable electricity are meaningless oversimplifications of real systems.

4.4 Nuclear Today

As with much data, accurate numbers can be hard to verify and depend on the source. If we look at data from 3 years ago, nothing much has changed, as we will see in a later section when reviewing changes in electricity production over the past 20 years.

Indicator	Year to Year	Starts	Closures
Startup / Closures	2023	5	5
Net Loss	2023	1.7GW	
Construction starts	2023	6 5 in China	nil
Last Decade (2014-2023)	2014-2023	6.1	6.7
Last 2 Decades	2004-2023	102	104
Where – China Peaked 2018 with 7	+2.3GW pa	49	nil
Rest of World		nil	51
Operable (2 decades)	2002-2023	438	413
Nuclear Share of electricity	1996-2023	17.50%	9.20%
SMR		none	

Figure 24 Nuclear Starts and Closures Summary

Figure 24 takes data from IEA 2003 report [63], GRS (Gesellschaft für Anlagen- und Reaktorsicherheit) [64] and the WNISE interactive data-wizard. [65]. In addition to these production units, there are an estimated 250 small reactors for research and pharmacy.

What the numbers tell us is that only China is building new plants but the contribution from nuclear to the electricity grid has fallen from 17.50% to less than 9% and is going to fall even more. Nuclear will be a rounding error in the contribution to electricity supply. In many countries it is zero. And likely to remain so.

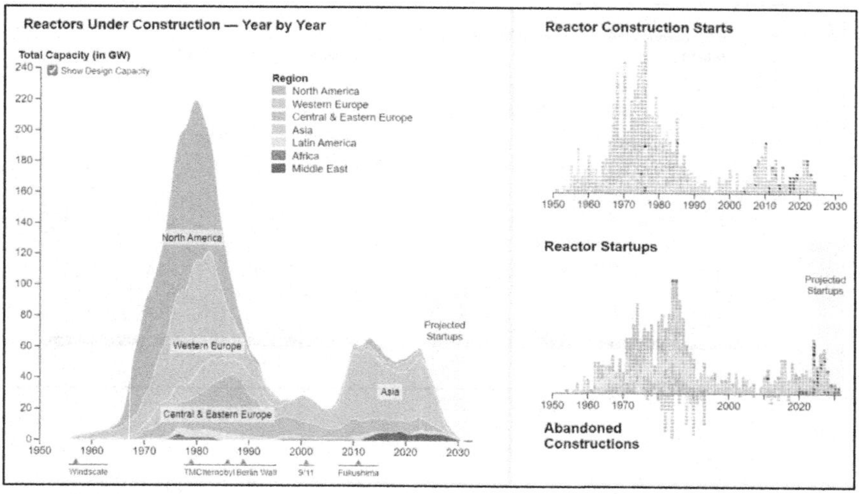

Figure 25 Numbers of reactors under construction, startups and abandoned from World Nuclear Status

What we can see clearly in Figure 25 is that nuclear power capacity in the USA and Europe increased from the 1960s through to 1995 to over 200 units. Then, building in USA and EU stopped. From 2010, construction began in Asia, mostly China. There has been no new major burst of construction, and furthermore, the projected startups scarcely show on the graphs.

4.4.1 No Change from 2000 to 2026

As pointed out in the rise and fall of nuclear, the Chornobyl and the Fukushima accidents put a huge dent in reactor starts. Larger units have replaced smaller retirees. Some is due to rethinking nuclear risks, and some is partly to the age of the plants. They became too expensive to continue. Too expensive to refurbish.

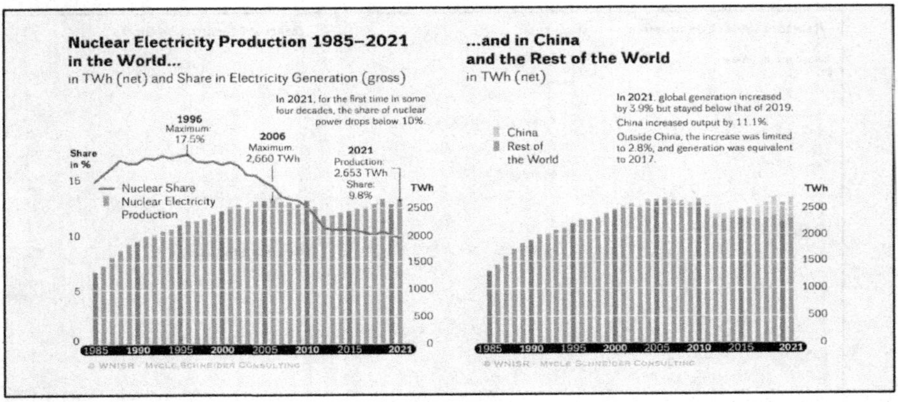

Figure 26 Nuclear electricity production 1985 to 2021

Figure 26 shows that generating capacity has remained stable over the past 2 decades. Any growth has primarily been due to updates of existing reactors. Had China not built new plants from 2015 to the present, the amount of electricity produced by nuclear would have declined and the share of electricity would have fallen even faster.

4.4.2 Nuclear Electricity Contribution is Falling

Every year, electricity demand grows as countries' GDP increases. Ember's 2025 numbers show this starkly.

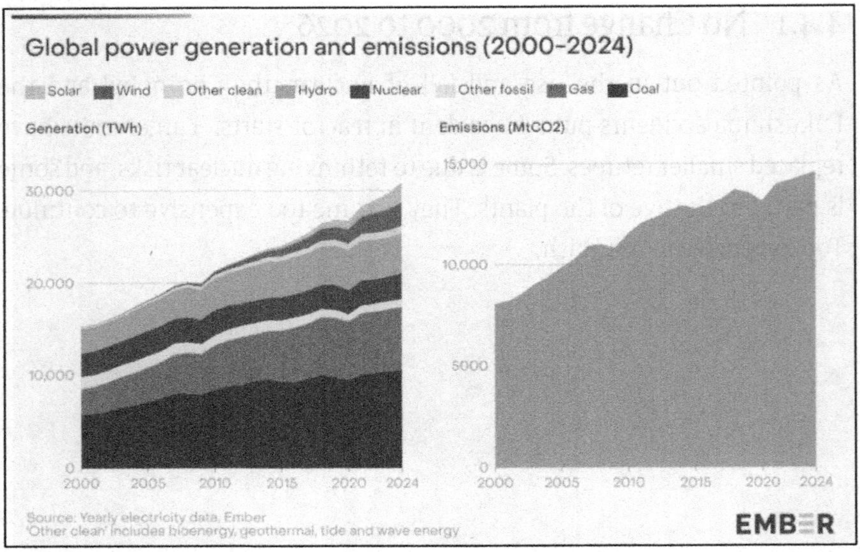

Figure 27 Global power generation and Emissions 2000-2024 - Ember

Figure 27 shows that electricity supply continues to rise, with coal and gas providing much of it until about 2015. In contrast, since 2015, the big increase in energy generation has been wind and solar. Some hydroelectric plants have been built in Asia, and none have been retired.

The opportunity for nuclear to increase its market share is remote. Solar, in particular, is made in a factory and assembled on site. In contrast, nuclear is made on-site. See further discussion about small modular reactors that are intended to be manufactured in a factory.

4.4.3 Global Energy Demand

A broader view is to look at the world's total energy demand and how it is supplied. The world uses about 166,000 TWh of energy. About 30,00 TWh of that is electricity generation, and about 20,000 is clean, the rest from coal and gas. We know about 65% of that primary energy is wasted. With electrified transport, the requirement would be 5 times less because electric vehicles are 5 times more efficient than fossil-powered transport. Likewise, in heating, where heat pumps are 3 to 4 times more efficient, we need much less electricity than when using gas for heating. The additional electricity to replace all of that fossil energy is about 40,000 TWh. Therefore, total electricity demand will be about 60,000 TWh versus the 166,000 today.

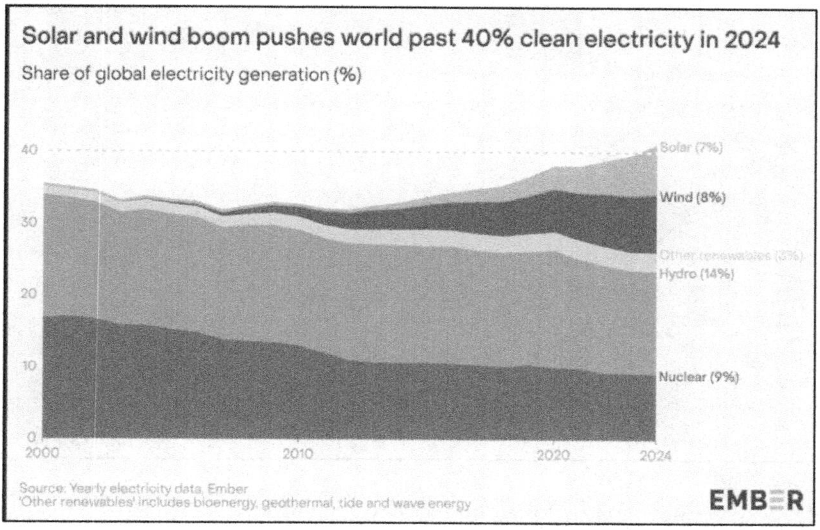

Figure 28 Share of Wind, solar, hydro and nuclear clean electricity from Ember 2024

Figure 28 [66], [67] shows the contribution of each of the clean energy technologies since 2000. Note the remarkable growth of wind and solar over the past 5 years. Wind and solar have now pushed past 40% of total electricity generation of about 30,000 TWh. Clean energy now produces more electricity than fossil fuel power plants. The contribution from each of the clean technologies is:

- Hydro= 14%
- Wind = 8%
- Solar = 7%
- Nuclear =9%
- Other = 3%

4.4.4 Solar Meets Increase in Electricity Demand

Predicting forward over the next decade? Solar will continue to explode, especially in light of the Middle East conflict in early 2026.

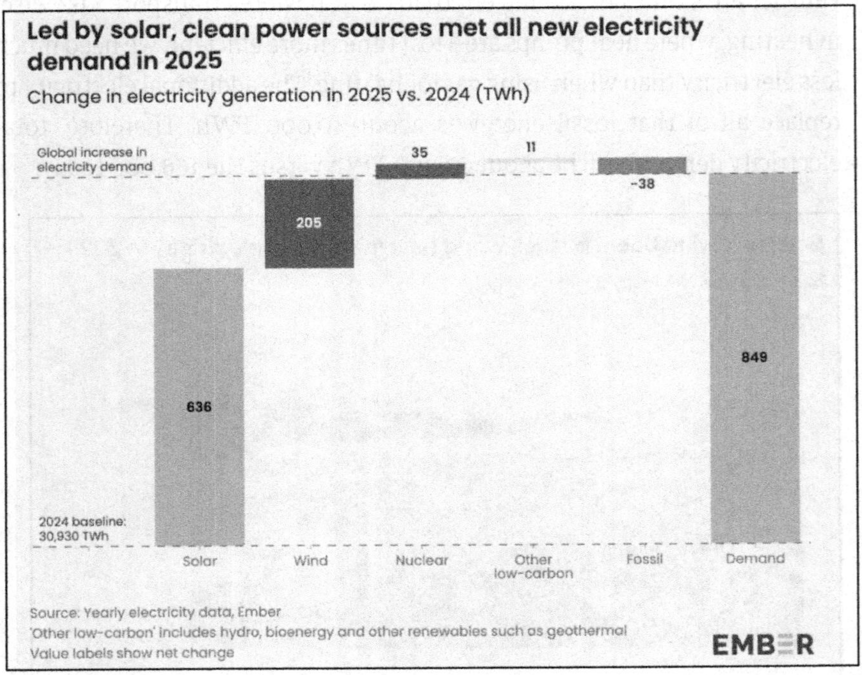

Figure 29 Global increase in Electricity Demand 2025 - Ember 2026

Figure 29 from Ember 2025 shows that solar production capacity is over 600

GW per year, and production capacity of solar panels is nearly 1 TW of capacity a year. If all that 1,000 GW of solar was actually deployed it would add over 1,500 TWh of electricity per year. Production capacity shows no sign of slowing and could rise to over 2 TW within 5 years. That adds 3,000 TWh per year. It means in just 6 years, there will be more solar than all of the current global electricity.

Where does nuclear fit? New nuclear supply is offset by nuclear retirement.

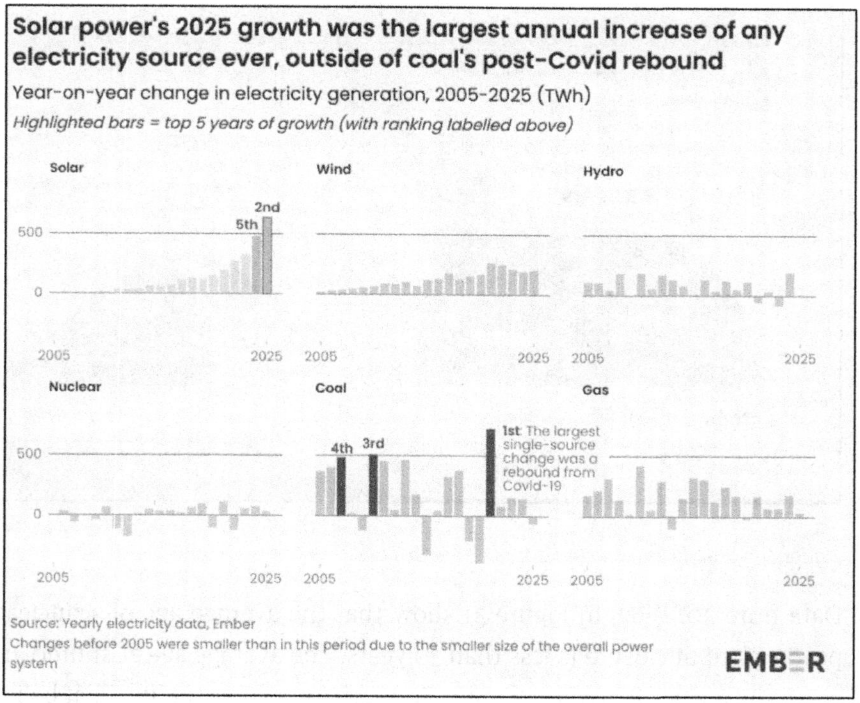

Figure 30 Annual increase and decrease of Energy sources 2015-2025 - Ember 2026

Figure 30 shows changes in electricity supply over the past 2 decades. [66] It shows that nuclear has not changed much, and retirements and new plants amount to less than 30 TWh per year. In contrast 2025, wind and solar added about 900 TWh to meet the extra demand for electricity. Moreover, the installation of wind and solar is pushing fossil fuel electricity generation out. It is now cheaper to build solar and battery than to supply gas or coal to a power plant. Nuclear deployment would have to increase 30 times more – and that simply won't happen.

4.5 Age of Plants

One of the myths about nuclear is the perception that nuclear plants would last for 80 years. It is an unsubstantiated claim.

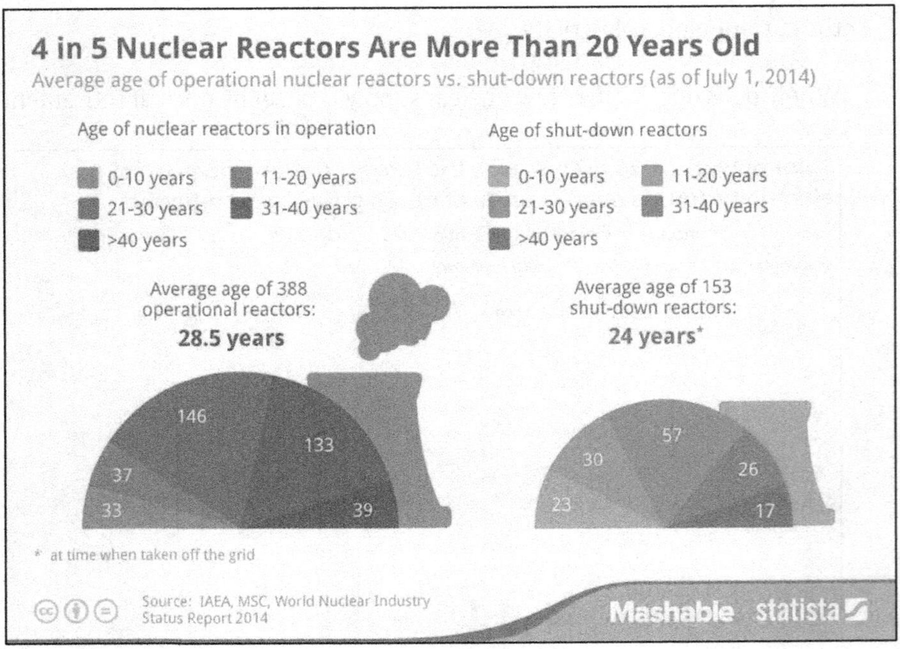

Figure 31 Age of nuclear reactors 2014 from Statista

Data from 2014 [62] in Figure 31 show that the average age of a nuclear power plant at closure is less than 30 years. The average age at shutdown was just 24 years for over 153 units. So much for an 80-year life. That belief is divorced from reality. The data is slightly different by 2024: USA average age = 41.2 (119 reactors), Asia = 17.4 (124), Europe = 35.6 (168), and Africa = 17.1 (2). As we will review later, these plants all have huge subsidies. Subsidies to extend their life are so large that they dwarf the cost of operation. Without subsidies, any plant up for refurbishment would be decommissioned. More details on subsidies are in a later section.

4.6 Future Plants

IEA projects a steady state over the next 2 decades[68] Nuclear energy, with around 413 gigawatts (GW) of capacity operating in 32 countries, contributes to both goals by avoiding 1.5 gigatonnes (Gt) of global emissions and 180 billion cubic meters (bcm) of global gas demand a year. It provides less than 9% of global electricity, and its contribution is falling with the new build of solar and batteries. It has not changed much over the past 20 years. It takes 14 years or more to bring new plants online. Costs and construction lead times will keep the number of new nuclear constrained through to at least 2035.

Site	Where	Size GW	Cost (b)	Cost $USb	$ per GW	Years
Flamanville	France	1.6	€13.30	14	9.0	12
Hinkley C	UK	3.2	£33b	42	13.1	22
Vogle, GA	USA	2.5	$35b	35	14.0	15
Barakah	UAE	5.6	$24b	24	4.3	12
Olkiluoto	Finland	1.6	$16b	16	10.0	18
Saeul 3/4	Korean	2.8	$9.30	9.3	3.3	6
Shin Hanul	Korean	2.8	$9.21	9.2	3.3	10?
NB. Most costs are "rubbery". Many are "expressed in 2016 /2012 terms" or similar. Years does not include planning and approval which add 3 to 8 years						

Figure 32 Construction costs & duration for 7 western nuclear plants.

In Figure 32 the construction costs and times for all Western reactors since 2000 are shown. The average cost is over $10 billion per GW, with construction taking over 12 years.

If we look longer term over decades 2020, 2030, 2040 as the following information from IEA and others, the overall projection is that new builds will just keep up with retirements. All of this will be predicated on political decisions, as all the investment and capital comes from the government. They can and do ignore technology and economics.

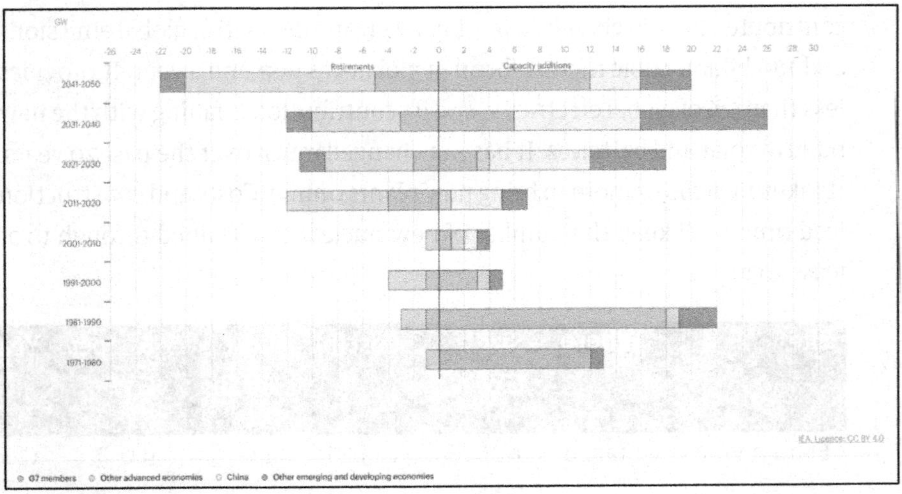

Figure 33 Nuclear power capacity additions and retirements in selected countries and regions by decade in the Net Zero Scenario by IEA

Figure 33 from IEA shows capacity will remain the same over the next decades with builds and retirements.

- In the 2020s, build 18GW; retire 11GW.
- In the 2030s, build 26GW; retire 12GW.
- In the 2040s, build 20GW; retire 22GW.

Even if a major burst of activity were to be announced today, the needle would not move until the middle of the 2030s. That's assuming new builds use proven Korean APR1400s and it also assumes approvals and planning take zero time. Adding approvals takes an addition 3 to 8 years.

4.6.1 Hinkley C

Hinkley C is on the Somerset Coast of England on a site of retired nuclear plants. The cooling water is taken and returned over 2 km offshore. The site is 14m above sea level and has a sea wall of 13m, and is reputed to generate enough electricity to supply some six million homes. When we look at that claim – it is clearly false. Ofgem estimates the typical household in England, Scotland and Wales uses 2,700 kWh of electricity and 11,500 kWh of gas in a year.[69]

- Generation capacity of 3200 MW.

- Production = 3200 * 365 *24* 90% = 25,228,800 MWh per year.
- Six million houses = 4.2MWh per year = 11.5 kWh per day.
- The average Australian household uses more than 20 kWh per day, and Canadians double that.
- The claim of the number of houses supplied by Hinkley C is inflated by 2 times, and moreover, does nothing to eliminate the use of gas in the house.
- At a price twice to three times that of wind or solar, the cost of electricity is on par with the cost of gas. That delays the transition from fossil fuels to gas and is discussed in detail in Part 1, "The End of the Fossil Fuel Era."

Rather than providing cheaper energy, Hinkley will increase electricity prices, slowing the change from gas. So, despite all the positive PR from the UK Government, it is clear that Hinkley C is a disaster in waiting for UK household electricity bills.

Planning for Hinkley C was started in 2007 for a capital cost of £8 billion with a planned cost of £100 per MWh. Negotiations with EDF in 2013 for construction led to the start of construction by 2017, with a price increase to £14 billion and an anticipated completion date by 2025. The roof was put on in 2023, with the price then at £33 billion and completion by 2028. The strike price for electricity had risen to £150/MWh for wholesale pricing.

By 2026, the cost is predicted to be more than £46 billion and completion by 2031. [70] Meanwhile, the UK government has said it wants nuclear to provide up to 25% of the UK's electricity needs by 2050 as part of its plans to combat climate change. This is nothing more than politics overriding common sense – technology and economics. The Government appears to be operating on a sunk cost fallacy. Putting good money after bad, and the £15 billion still to be spent, would be more appropriate. They could do the following with £15 billion.

- 10 GW of wind in the North Sea with 5 GW of batteries in the South of England.
- 750,000 houses provided with a grant of £20,000 to install solar, battery and a heat pump to homeowners and public housing. The energy bill after retrofit will be zero.

4.6.2 Sizewell C

The Sizewell C new nuclear plant is situated in UK, on an existing site. Sizewell A was decommissioned in 2005, and Sizewell B is still operating, and decommissioning is planned for 2035. Sizewell C is planned to be a copy of Hinkley and built by EDF and the China General Nuclear Power Group. The UK claims construction will be easier following Hinkley C, but that is unfounded optimism. Details of the project in summary are:

- Sizewell C = 3.2 GW
- Begun in 2012, 13 years ago and is slowly progressing with successive governments being hooked into the project.
- Costs are escalating.
- In 2020 the estimated cost was £20 billion.
- By 2025 and according to the Financial Times, the cost is now expected to double to £40 billion, or $79 billion. [71]
- Taxpayers' money £19 billion (48%).
- The UK Govt has committed to £2.1 billion in contracts already.
- The development is a Regulated Asset Base Model where bill payers will need to pay at least £1 per month – that's £240 over 20 years plus a surcharge on every bill.
- The latest projected final cost if built by 2040 will be over £60 billion ($AU123 billion).

The project secured funding because the UK Govt, with 48% of the deal, and a number of banks took up the balance of the first £5 billion funding round. How the balance of the project costs will be funded is not determined. Belief systems can override economics. In a statement, Sizewell C's co-managing directors, Julia Pyke and Nigel Cann, said the claims about Sizewell C's costs were *"not accurate and do not reflect the significant savings we are already making"* following its work on Hinkley Point C. The pair went onto say:

"The real question to ask is what is the cost of not doing Sizewell C. This winter we have seen prolonged periods of dull, calm weather and low output from wind and solar. At times, gas provided more than 60% of our power needs, pushing up electricity prices and adding to our carbon emissions. A low-carbon future needs new British nuclear, for lower electricity costs, energy security and thousands of great jobs which will help to transform communities across the country."

These comments continue the public lie that renewable energy is "unreliable". They continue, like other nuclear advocates, the false assertion that electricity grids need "baseload". More of that in Part 1: the end of the fossil fuel era.

4.7 Nuclear Technology

Technology has not really changed substantially since 1950's - 70 years. While one would expect mature technology, there is little sign, particularly with dealing with waste or improving safety. There have been some new designs – the most recent is a Generation III+ pressurised water reactor (PWR) design and the European Pressurised Reactor (EPR). South Korea has settled on a PWR1400 design. The most recent Chinese are Hualong One, CAP-1000 and CAP-1400. They follow standard designs as compliance is extremely costly. Warships still use reactors not much different from the first reactors in the 1950s. Improvements include safety and efficiencies. The most significant improvement has been the refurbishment of existing plants and the increase in their output. We need to differentiate between nuclear for weapons, nuclear for power generation and nuclear for medicine. Nuclear moved from weapons of war to weapons for global suicide and we are 90 seconds from midnight (Nuclear Armageddon). [72] China, Russia, and the United States are all spending huge sums to expand or modernise their nuclear arsenals, adding to the ever-present danger of nuclear war through mistake or miscalculation. Even Israel and Iran are in on the act. Proponents of nuclear power are quick to dismiss the notion that nuclear is a stepping stone to nuclear weapons, but government documentation is clear.

Nuclear medicine makes sense, although new medical genetic techniques may well surpass some medical nuclear technologies. A recent development of a small nuclear battery in a button battery to power medical devices is novel, but it ignores advances in other technologies. [73]. Most support nuclear research for either medical research, or research to deal with the vast reserve of nuclear waste from medicine, weapons and retired nuclear power plants.

4.7.1 Low Land Requirement for Nuclear

Nuclear advocates like to point out that nuclear power does not require land and is renewable. What is often forgotten about nuclear power is the extensive mining and enrichment needed to fuel the plant, and U-235 is used up, and has to be replenished. It is not renewable.

A 1,000 MWe pressurised water reactor requires each year.

- 27 tonnes of uranium.
- 18 million fuel pellets.
- 50,000 fuel rods to house pellets.
- Water to cool the plant.

In contrast to other electricity generation the nuclear power station itself is dense and does not need large tracts of land.

- Coal power: An equivalent size requires 2.5 million tonnes of coal and all the associated coal mines and transport systems, and waste.
- Wind farm: With 250 8MW turbines. A claim that it needs thousands of hectares is false. Each turbine is spaced to occupy 110 ha for wind efficiency, but in practice, a wind turbine occupies less than 100 m^2. The area actually used by turbines is less than 1% of a wind farm, and most of that is roading and access. The rest of the land can be farmed or left in forest. So a 1 GW wind farm needs less than 1 ha actual area.
- Utility Solar: A 1 GW solar farm is about 3 million panels and 2,000 ha of land. In many cases, agronomics, where solar panels are paired with either livestock or crops beneath, is more productive than land itself.
- Rooftop Solar: The whole of the worlds energy demand could be electrified with just half of the rooftops with solar. They aren't now because utility solar is cheaper than rooftop solar. The primary cost is labour costs for installation. No new land needed.
- Gas plants: Gas plants are dense (20ha) but if the gas fields and pipelines are accounted for then they have a large footprint.

As we discussed in Part 1, those wind turbines could be built in a quarter of the time, and electricity would be 1/3rd the cost.

4.7.2 Nuclear is Low Emission

Nuclear power is low-emissions, but not that low due to energy-ointensive mining, refining, and enrichment. Little to no Uranium is recycled or repurposed - as we will discuss later. The measure of emissions is calculated using a life cycle analysis (LCA), which accounts for emissions from mining the raw materials all the way through construction, operation, and end-of-life. It is complicated with large projects. The LCA of various technologies range from:

- Coal = 800-1,000 grams of CO_2 per kilowatt hour (gCO_2/kWh)
- Gas = 400-500
- Solar = 38
- Nuclear = 15
- Wind = 8

Generally, the LCA of nuclear power is thought to be similar to that of wind, although a recent LCA study of European nuclear power [68] concluded that the results for a new nuclear power plant in Europe were higher than generally reported. Three methods were used: process-based, input-output, and hybrid life cycle assessment. The results range from 8 to 64 gCO_2/kWh. The averages for the three methods are 16.9, 24.9 and 27.6 gCO_2/kWh. These are either well above or at the upper end of the range of possibilities (5 to 22 gCO_2/kWh) stated in a report for the UK's Committee on Climate Change, and significantly higher than the median value of 12 gCO_2/kWh presented by the Intergovernmental Panel on Climate Change. They are also higher than the values acknowledged by the nuclear industry.

In contrast, wind and solar are falling due to lower emissions from mining and electricity used in manufacturing processes.

4.7.3 Reliability and Capacity Factors

There is much misinformation about the reliability of nuclear plants. They are the pinnacle of engineering, costing billions with safety systems, and in general, are very reliable. Like all thermal electricity generation, nuclear transforms uranium fuel into heat. Heat then creates steam (often at extremely high temperatures and pressure). That steam drives turbines. Steam and turbines need maintenance, whether the heat source is fossil fuel or nuclear. The power plant needs cooling, and most of it comes from rivers or the ocean. If there is not enough water, the plant cannot be cooled, and then it shuts down. Korea, with modern plants, has about an 80% capacity factor. USA 90%. They can be as low as 50% in some countries. For a full understanding of capacity factors, read more in Part 1. The Finnish plant was launched with great fanfare, but 12 months later, it was shut down for repairs. [74], [75] The shutdown in Finland highlights a risk with all large electricity generation systems. If you drop 3 GW of generation, what takes over? With the failure of a single wind turbine, one might lose 8MW. These large generators, coal, gas or nuclear power plants need backup systems, and in most electricity grid markets, there are idle generators providing backup services. They cannot rapidly scale up or down to meet fluctuations in demand. Read more about the complexity of grid systems in Part 1 and how part of the cost of electricity is for these FCAS services (Frequency Control and Auxiliary Services).

4.7.4 Mining of Uranium

The availability of uranium drives the future of the nuclear energy industry; an upper limit on uranium prices will determine whether it is economically viable to continue utilising the present generation of light water reactors (LWRs). The alternative is a move towards fast reactors, which use less uranium than LWRs; however, they require substantial investment and development, and might never become economically viable unless uranium prices rise considerably. Break-even mining requires an ore concentration of about 340 ppm. Concentrations below 100 ppm require too much energy, even with more energy-efficient mining. Researchers estimate that between 80 and 100 years of economic mine life remain. [76] [77] That is not much more than fossil fuel of 50 years. In contrast, we have a billion-year warranty from the Sun for solar. As high-grade uranium ores become depleted, the trend is to shift toward ores with lower uranium concentrations. The energy required to mine and mill lower-grade uranium ore increases. Low-grade ores of any mineral [77] require more energy. A typical figure for total energy consumed in mining uranium is 0.2 gigajoules

(3.6 MWh) per kilogram.

Rising demand for uranium and diminishing availability of high-grade ores will inevitably cause uranium, especially U_3O_8, prices to rise. If uranium prices rose high enough, they would push up the operating costs of LWRs to the point where a closed fuel cycle and/or thorium fuel could compete economically. The Levelised Cost of Energy (LCOE) is the cost per unit energy at which a particular power station will break even over its entire lifetime.

While terrestrial uranium resources are limited, there is over 4.5 billion tonnes of uranium in seawater. It is only at concentrations of 3.3 ppb but researchers have achieved selective extraction. Sea reserves are 1,000 times more than terrestrial reserves. Two problems face this source of uranium. (a) Cost. Initial estimates confirm high cost. (b) Can that process happen at the scale needed? The extremely low uranium concentration and the presence of large amounts of coexisting ions, especially K+, Na+, Ca_2+, and Mg_2+, make extracting uranium from seawater very challenging.

4.7.5 Production of Uranium Yellowcake

Conventional mining (e.g. open pit or underground) has been the main source of uranium. The mined ore is crushed in a mill. Water is added to produce a slurry of fine ore particles and other materials. The resultant slurry is leached with sulfuric acid or an alkaline solution to dissolve the uranium. The remaining rock and other minerals are undissolved.

In contrast in-situ leaching method is now half of the world's uranium mining. There is no major ground disturbance. Water is injected with oxygen or with an alkali, acid or other oxidising solution which is circulated through the uranium ore, extracting the uranium. The uranium solution is pumped to the surface and then the solution is separated, filtered and dried to produce uranium oxide concentrate, often referred to as 'yellowcake'.

4.7.6 Uranium Enrichment

The yellowcake has to be enriched to take the U-235 from 0.7% to 3% to 5% to be used. Enrichment of uranium is a complex and energy-intensive process. [78]

- Low Enriched Uranium (LEU) is the basic material to fabricate the nuclear fuel rods or pellets for conventional nuclear reactors. It consists of uranium hexafluoride, a white-grey, waxy solid at standard temperature and pressure with a content of 3% to 5% of U-235. The maximum content is 20% uranium. [79] Some LEU is made using old nuclear weapons and diluting down the U-235.
- Highly enriched uranium (HEU) has a 20% or higher concentration of U-235 and is essential for certain specialised reactor designs.
- Weapons grade uranium is over 60% enriched and at that level you can make bombs. Most bombs are at 95% enrichment, but that is more a function of the warhead's weight. The political stoush with Iran is that they have enriched about 500kg to a 60% level and were only weeks away from making a dozen atomic bombs. Clearly, Israelis and most nuclear-weapon-possessing countries don't want that.

Who has control of enrichment has control of the industry.

Company	2020	% Capacity	2025	2030
CNNC	6300	10%	11,000	17,000
Orano	7500	12%	7500	7500
Rosatom	27,700	46%	26,200	24,800
Urenco	18,600	31%	17,300	16,300
Other	66	0%	375	525
Total	60,166	100%	62,375	66,125

Figure 34 Enrichment companies globally demonstrate 46% of nuclear fuel is sourced from Russian-controlled entities.

As Figure 34 shows, over 46% of enrichment occurs in Russian-controlled businesses in the 'Stans, and the US gets over 25% of the uranium in its power plants from Russia. When the USA started de-arming nuclear warheads post-Cold War, they took nuclear material from the stockpile of bombs to be scrapped, which were at 95% enrichment and reduced them to 5%. Uranium enrichment factories shuttered. Currently, the USA is pursuing political moves to fund the construction of new enrichment facilities, and while they may have a couple of years' supply, building an enrichment facility will take time. Lots of money. Government subsidies again.

Meanwhile, building a new nuclear plant in a country without an enrichment facility means being held hostage to Russian interests. As Europe discovered to its horror, being hostage to Russian interests for gas did not turn out well.

4.8 Small Modular Reactors (SMR)

Small modular nuclear power plants or nuclear SMR reactors have been talked about for the past decade to overcome the issue with conventional large nuclear. SMR proponents argue that manufacturing in a consistent design in factories will be cheaper than bespoke large systems. However, there are over 17 different types of SMR designs, reactors, and there are still no demonstration units. Recent reports continue to say they are still too expensive and too risky, but that has not deterred Microsoft and Amazon from stumping up money to get some of these producers to produce. [80]. [81] SMR have the same issues as large reactors.

- Waste not addressed.
- Absence of social license or willingness to accept a nuclear reactor in your backyard.
- 18 different designs.
- 37 variants.
- The first one by 2029 - but NuScale's failure to secure funding puts this out to the 2030s.
- Under 300 MW in total.
- More toxic.
- Cost reduction by "factory line" manufacturing is overhyped.

SMR technology was investigated in the 1950s and 60s. The technology is much the same today as it was half a century ago. In engineering, the question is how many iterations it takes to get economies of scale in manufacturing. Economies of scale can be both internal and external. Internal economies are caused by factors within a single company, while external factors affect the entire industry. Generally, it is a U-shaped curve. In most engineering, one needs to build a dozen or so prototypes. A rule of thumb is that if one manufacturer handles all manufacturing, doubling production would reduce costs by about 10%. Another measure is that you would need to build at least two to three times as many nuclear plants as there are currently. That is about 700 units to get costs down by 50%. [82]

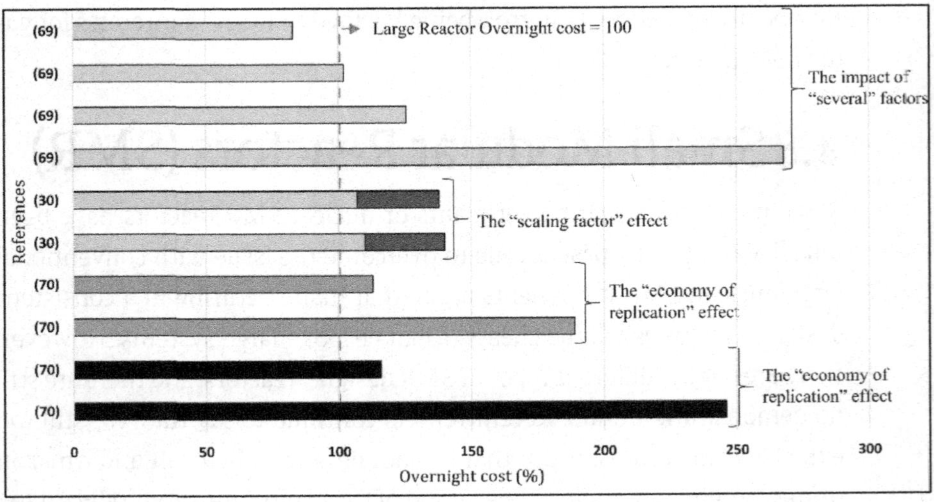

Figure 35 Overnight construction costs of various nuclear power plant units from Mignacca & Locatelli

Figure 35 shows the overnight cost of SMR versus large reactors (LR) from various analyses globally. The LR is shown at 100, but several factors, as described in the costs of nuclear section, show costs range from $8 to $22 billion per GW. In contrast, SMR have overnight costs that vary from 20% more expensive to 2.5 times more expensive. At this stage in the design, modular units, made under factory conditions, will be more expensive. While advocates for SMR promote them, the reality may be quite different.

- Transportable – about a container size. That would only be for very small units – there are none in prototype phase.
- Claimed that 4 to 12 units each of 50MW would allow larger systems. Most close to commercialisation are closer to 300 MW.

- Provide power to smaller communities.
- SMR will compete with wind, solar batteries, and small engines.
- Note that some claimed SMR sites, such as Ontario 300 MW nuclear reactors are not an SMR by definition.

As highlighted by the terms "Small" and "Modular", SMRs present three main peculiarities compared with large-scale traditional reactors: smaller size, modularisation, and modularity. What is clear today is that they are dirtier, more expensive and less competitive with alternative electricity solutions. Bundled packages of wind, solar, battery and diesel generators is already operational in remote communities and mine sites. They are much cheaper than full-time diesel generators and have demonstrated they cut diesel and gas by 90%. They have faster deployment and are readily available today.

4.8.1 SMR - Oklo

Conventional reactors use moderators like water to slow neutrons down before they split, or fission, uranium and plutonium (Pu) atoms. Oklo's reactors won't use moderators, enabling the construction of much smaller plants and allowing neutrons to move faster. Faster-moving neutrons can sustain nuclear fission with a different type of fuel. Oklo's first reactor will generate 1.5 MW using HLEU with 5%-20% U-235 from used fuel recovered from an experimental reactor at the Idaho National Laboratory that was shut down in 1994. Oklo says the plant will run for 20 years without refuelling, but the long-term objective is for plants to run for 40 years without refuelling.

4.8.2 SMR - TerraPower and Natrium

Terrapower has three different technologies. Bill Gates-funded TerraPower is building a $4 billion advanced nuclear plant in Wyoming with government funding ($2 billion) plus private equity. They have just raised an additional $600 million in a round for their first plant [83] using Natrium technology. It features a 345 MW sodium-cooled fast reactor using high-assay, low-enriched uranium fuel, with a molten salt-based energy storage system that can boost output to 500 MWe for more than five and a half hours when needed. They are currently building the sodium storage and generator part of the plant while waiting for the nuclear part. They intend to open its Natrium plant in 2028.

4.8.3 SMR - TerraPower MCFF

TerraPower's Molten Chloride Fast Reactor (MCFR) is a molten salt reactor (MSR) that uses liquid chloride salts as both fuel and coolant, which circulate through the reactor core to generate heat. Salts serve as both the reactor's coolant and fuel. The MCFR design specifically requires molten chloride salt, which allows for fast-spectrum operation. In the fast neutron spectrum, neutrons are not slowed down (e.g. by colliding with water or graphite) and move very quickly, making the fission reaction more efficient. In MCFR cores, nuclear fission occurs and heats the fuel salts directly. The MCFR then distributes heat from the molten fuel salt through a heat exchanger to an inert salt in a second loop. Heat from the non-nuclear secondary loop is then safely used for electricity generation, process heat or thermal storage. Core Power is using similar designs.

4.8.4 SMR - Terrapower Travelling Wave Reactor

Terrapower is also proposing an innovative travelling wave reactor (TWR) design. In the centre of the core sit rods of enriched U-235, surrounded by rods of depleted or natural uranium (U-238). The U-235 serves as an initiator, kickstarting the travelling wave reaction – a slow-moving chain reaction of concentric waves of fission. The travelling wave reaction will then slowly convert the U-238 (called depleted uranium) to fissionable plutonium. Periodically, to sustain the fission reaction, the in-vessel fuel handling machine shuffles the fuel, swapping expired fuel rods from the centre of the core for fresh fuel rods from the outer edge. Most reactors in use today need to be refuelled every 18 to 24 months. Terrapower says the MCFR can be fuelled as part of a continuous process that does not involve extended shutdowns, opening the reactor vessel, or moving numerous fuel assemblies.

4.8.5 SMR - Rolls-Royce

Rolls-Royce has been servicing the British Navy nuclear for decades with the largest workforce in the UK. They have just recently announced they are going to re-enter the market for SMR nuclear. They say they are "walk away safe" for emergency shutdown. A meltdown or explosion is impossible. The size is equivalent to two football pitches, built on a seismic raft and can be built inside a canopy. They claim they can be built within 4 years, rather than 20 years and will be good for 20 years. The reactor itself will be 16m long, 4m in diameter, with a 417 MWh LCOE of $85 MWh. The UK Govt tipped in £250 million in 2022, and Rolls-Royce was hoping to have the first site in 2032 and have another 16 at about £2.5 billion each. [83] In June 2025, the UK Gov committed £14.2bn (€16.85bn, $19.26bn) investment in the Sizewell C nuclear power station. Rolls-Royce has already agreed to produce an SMR in the Czech Republic, and it is in the final two in a competition in Sweden. In this nuclear announcement in June 2025, Rolls-Royce won a competition to be the first company to try to build small modular nuclear reactors (SMRs), including potentially at Sizewell. [83] A separate government release said SMRs would collectively generate up to 1.5 GW of electricity, suggesting that Rolls-Royce will be granted permission to build at least three SMRs, although it did not reveal the locations of the first UK SMRs, which some in the industry had hoped for to speed the process along. They are likely to be sited beside retired nuclear power stations such as Oldbury in Gloucestershire and Wylfa in north Wales, or at Sizewell. Rolls-Royce beat competition from the US companies Holtec and GE Hitachi, while the Canadian-owned Westinghouse dropped out earlier. It's the same story; nuclear power is only built with government money.

4.8.6 SMR - Radiant

Radiant Nuclear is developing a microreactor that they call the "world's first portable, zero-emissions power source." The idea for the microreactor, called Kaleidos, stems from CEO Doug Bernauer's time working at SpaceX, where he was tasked with devising methods for producing enough power on Mars to allow future colonies to survive and also travel between Earth and the red planet. The high-temperature gas-cooled reactor is designed to produce 1.2 MW and operate for 5 or more years before refuelling. They are focused on remote communities, disaster areas and military bases. They have raised some money and got to the experimental design and front-end stage. Don't expect much more until 2030.

4.8.7 Micro Nuclear Reactors

Micro-reactors are also in the news, and it seems there are plenty of investors willing to front up for designs and government subsidies. Most are 1 MW to 5 MW capacity. Radiant, Ultra Safe Nuclear Corporation and Westinghouse have been awarded US Department of Energy funding totalling $3.9 million for front-end engineering and experiment design. The test bed facility is at Idaho National Laboratory.

- A Microreactor project at Idaho National lab is doing research on a tiny, relatively inexpensive nuclear reactor. It's more of a nuclear battery, than an SMR, and is in the range of 1 to 5MW. Its use would be in places like Alaska or remote communities without access to wind or solar.
- The first prototype MARVEL reactor (Microreactor Applications Research Validation and Evaluation) and target date for operation in late 2023, making it the first advanced microreactor in the United States.
- CNBC reported in Jun 2022 a single microreactor could power a small community. A hospital or remote military base of 1,000 to 10,000 is a small community. Ultra Safe Nuclear Corporation (USNC), based in Seattle, Washington, is in last-ditch attempts to stave off collapse [84].
- Westinghouse achieved a crucial milestone in microreactor development with their eVinci design. Concerns about nuclear waste and safety persist. When? Anyone's guess at this stage.

4.8.8 SMR Conclusion - Too Many, Too Late

Liebriech points out the fantasy [85] and says that in comparison, in 2024, wind and solar generated 4,600 TWh of electricity. To match that amount of power with the 300MW SMRs being designed by GE-Hitachi or Westinghouse, you would need to build 1,960 of them, or 1,250 Rolls Royce SMRs, each producing 470MW, or over 8,000 smaller units like those promised by Oklo and Kairos. Given that these take 25 years to build, by 2025, you will be delivering less than 10% of the soaring global electricity supply, and meeting less than 5% of total energy demand.

Meanwhile, wind, solar, and batteries, at 10% annual growth, will have replaced all of those 166,000 TWh of primary energy, supplying over 65,000 TWh of electricity. To make any dent in global emissions, you would need to build not just thousands, but tens of thousands of SMRs – each of which requires permits, customers, fuel, security and trained staff. And still have nowhere to store the waste. Or have a few blowups killing thousands of people. We will be lucky to see half a dozen SMRs by 2030, and a few hundred by 2040.

> *Small modular reactors won't achieve economies of manufacturing scale, won't be faster to construct, forego efficiency of vertical scaling, won't be cheaper, aren't suitable for remote or brownfield coal sites, still face very large security costs, will still be costly and slow to decommission, and still require liability insurance caps. They don't solve any of the problems that they purport to while intentionally choosing to be less efficient than they could be. They've existed since the 1950s and they aren't any better now than they were then. Most of the attention and funding is misguided at best, and actively hostile to climate action at worst.*
>
> Michael Barnard 2021 *[86]*

And still no suggestion on how to deal with the waste.

4.9 Thorium Reactors

Thorium reactors sound better than a fairy tale. Unlimited renewable energy, available 24/7 at costs lower than firmed wind and solar. And it may be that Copenhagen Atomics [87] may have been smart enough to pull it off. Another decade will see. Thorium, like nuclear fusion, has been 20 years and $20 billion away from a real product. China commissioned a pilot thorium reactor in the Gobi Desert, which is running successfully, sometimes. Press releases say a pre-production engineering prototype is scheduled to come online by 2026 and will be commercially available for the first time by 2030. There is another problem. Thorium reactors can create U-233 which can be used to build atomic bombs. The USA shut down their thorium reactor in the 1960, so it has its advocates.

A golf ball of thorium costs about $100, and would supply all the energy an individual needs for their lifetime and the energy to support that - transportation, buildings, food. But the problem is, no machine can take all that energy from that ball. As we saw previously nuclear reserves are less than 80 years and we only use the U-235, which is just 0.7% of combined U-235 and U-238 ore. Thorium reactors use thorium and molten salt. The salt becomes the fuel. Thorium is a widely available product. The absence of water means that it avoids a situation where a meltdown is a risk.

Feature	Solid Fuel Reactor	CA Thorium Reactor
Expected Plant Life	60+ years + extension	50+yrs + extension
Price per kg fuel (U/Th)	$4500 (5% enriched)	$50 *
Fuel used	U-235	F_7LiTh
Energy Output per kg fuel	1-2 GWh (Thermal)	22 GWh (Thermal)
Construction Time for 1GW	15 years	6-18 months
Modularity	No	40' Container 100MWh
Energy Served to Customer	Electricity	5600C
Electricity Price MWh(e)	$60-$120	$20-$40
Build time from 2024	2034	2028
Note. Thorium weight capex also includes heavy water $F_7LiThPu$, cocoon, kickstarter fuel and other elements not consumed		

Figure 36 Comparison of nuclear reactor with CA Thorium by CA2024

Copenhagen Atomics [87] with Figure 36 are claiming they could build one fully functioning unit per day - 100 MWh per day, and deliver a temperature at 560°C. The moderator is heavy water. They claim the major challenge would be deployment of the units - assembly on site and turbines, etc., to use that energy. Copenhagen is at a 1MW prototype phase (testing in 2026) and claims it will be at mass production by 2030. Thorium may be further developed to use nuclear waste and reduce its half-life from 100,000 to 300 years. Due to regulatory challenges, they have postponed the use of spent fuel in a fast-breeder thorium reactor to the mid 2030s and claim they may be able to extract 10 times more energy from uranium than from the original fission. The 50 cm thick containment is built in segments and put together on site, enabling factory production and assembly only on site. It may be the ultimate "SMR". Their business plan is to own and operate and provide heat and steam to run turbines or for commercial use. It seems a smart move to own and operate, given the shortage of skilled workers in this industry.

But like those developing SMR reactors, if Copenhagen Atomics can succeed, it's good news—even though their first 100 MW plant won't be ready until 2028. They could scale up. Producing one per day would amount to 36 GW annually. In comparison, 473 GW of solar was installed in 2023, and an estimated 1 TW in 2026. This number is growing rapidly. Even accounting for a 20% capacity factor for solar and a 90% capacity factor for thorium, it will still be too little, too late.

4.10 TRIGA Reactors

TRIGA (Training, Research, Isotopes, General Atomics) are from General Atomics [88] and is a swimming pool reactor that can be installed without a containment building. They are designed specifically for research and testing use by scientific institutions and universities for purposes such as undergraduate and graduate education, private commercial research, non-destructive testing and isotope production. I know it well, as I spent a couple of years during a PhD in a lab inside the Corvallis 1.1MW plant, constructed in 1967. Fuel for TRIGA is a low-enriched uranium zirconium hydride (UZrH) and is made by the French. A new company, Aalo, has raised funds to make container-sized plants for energy and electricity [89] using the design of the Idaho National Laboratory Marvel. The MARVEL is an 85kW microreactor using sodium-potassium-cooling. Aalo has a target for $30/MWh with a first unit by 2026. Ambitious? A 10MW plant would need 15 containers, so this is not a plan for GWs of energy.

4.11 Nuclear Fusion

With billions of dollars into research, is it fusion or fiction for nuclear power? Unlike fission, which takes a heavy molecule (Uranium) and splits it, fusion takes small molecules and combines them to make a heavier molecule.

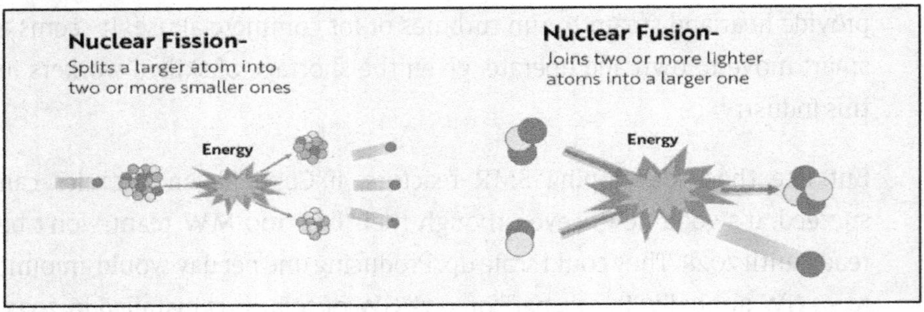

Figure 37 Nuclear fission versus nuclear fusion from Ark Invest

Figure 37 shows this difference. It is the basis of our universe, and physicists estimate that the first fusion stars appeared at least 100 million years after the Big Bang. Nuclear fission has been the subject of research since the 1940s and requires sophisticated engineering.

4.11.1 Technical Barriers to Fusion

The same cost challenges with fission also present themselves with fusion. Waste is less of an issue. Safety – as energy is needed to drive the process, there is no risk of runaway safety. With reduced radioactivity, social license is likely to be lower. The raw materials (water/air) reduce sovereignty risks. Construction delays are likely to be more severe, given the lack of a history of commercial plants. Challenges ahead include:

- The energy-out to energy-in ratio has to increase 1000 times more.
- Requires new technology to capture the energy to convert it to electricity.
- Fusion creates lots of high-energy nuclei, including fast neutrons, protons, deuterons, helium-3, helium-4, tritium, etc., which are enough to cause secondary nuclear reactions in the metal of the containment vessel. These neutron emissions render materials severely radioactive. In short, physicists have never seen a

proposed fusion route to zero radioactive byproducts. How to handle that waste?

- There is a whole supply chain needed for the enrichment and refining of hydrogen/helium or cobalt as fuel for fusion.

Globally, there are more than 80 research initiatives. [90] With three different strategies, the challenge is to extract more energy from fusion than is needed to initiate the process and achieve fusion. The tokamak is central to fusion. Magnetic fields control the plasma. These are engineering feats in themselves, with ship-sized engineering down to millimetre scale. Constructing an artificial star is difficult. Consider the engineering challenges that within a space of 2 metres, the temperature can go from outer space, 4°K, to 10 times hotter than the sun, 150 million °K. Tokamaks use 3 different methods. The tokamak is not the only design; there's also, in magnetically confined fusion, something called a stellarator, which kind of looks like Salvador Dali drew a doughnut. For a fuller explanation, check out Wikipedia. [90]

4.11.2 Timeframe to Fusion

There are good reasons why, after eighty years, we are still about "20" years away from sustained generation of net energy. There is a wide divergence of views about the timeframe, and if you are a private company raising money, the timeframe is "real soon". The best guess is that it will be the latter part of this century before we get electricity from a fusion reactor onto the grid. Constructing an artificial star is difficult and expensive.

The biggest fusion project on the planet is the International Thermonuclear Experimental Reactor (ITER) in France. It is behind schedule and over budget, ballooning from an initial estimate of $6.3 billion in 2006 to $22 billion in 2023 [91]. Started in 2013 and expected to have successful fission plasma by 2032. [92] Some say the budget could top €65 billion. It has not stopped private investors who have ploughed $1.3 billion in 2025 alone. And ITER does not plan to generate electricity, nor scale up to multiple units, whereas private companies need to get electricity or heat out to pay for the fusion.

Most research is to prove that you can get more energy out than put in.

Minimal energy is produced – enough to boil a kettle. Most fusion breakthroughs suffer from the same problems – they turn 300 MJ (megajoules) into 2 MJ and use those 2 to produce 3 MJ. Technically, that's a gain, if you ignore the massive loss beforehand.

Commonwealth Fusion Systems claims a fusion power plant will go live in the next decade and produce 400 MW. [93] Various research groups are working on the technology, and a recent USA pilot experiment produced 23 times more energy than the previous one.

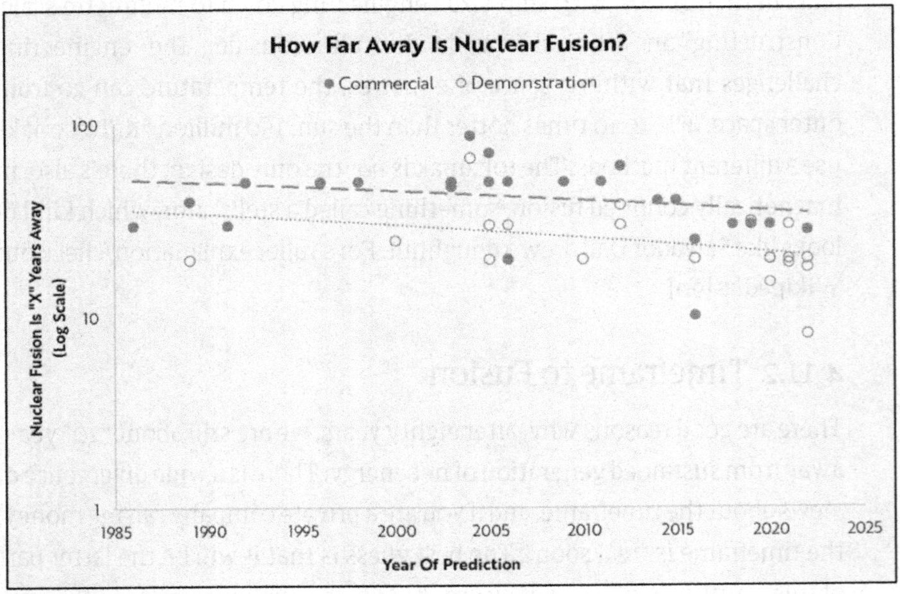

Figure 38 How far are we away from nuclear fusion? Ark Invest

Is it getting close? In Figure 38 Ark Invest shows the prediction year and in 1985 most predicted about 50 years away. The trend line shows it is still decades away although some optimistic people are saying a demonstration plant is under ten years with heat production still 30 years away.

Supporting research into new technologies is always good and should be encouraged. However, the dire consequences of climate change acceleration suggest that fusion research should only continue at the research level. We need to stop emissions today by electrifying everything and not in 50 years. Other funding should be diverted to energy technologies with short time frames and potential for widespread use. The barriers to fusion are the same as for fission: cost, waste, safety, social license, sovereignty, and construction delays.

4.12 Costs And Govt Subsidies

Nowhere around the world is the private sector saying, "*We want to build nuclear reactors and we will fund it.*" They only get built with direct government involvement – governments going guarantors on the loans, or at least guaranteeing a price for the electricity produced. Moreover, as an aside, note that it is quite an interesting philosophical position for national fiscally conservative political parties in many countries to take: they want to build it rather than leave it to the free market. Governments subsidise nuclear power. The Manhattan Project, which developed the atomic bomb, employed more than 130,000 people and cost nearly US$2 billion (equivalent to US$23 billion in 2007 dollars). Governments hide the extent of subsidies, and it is the taxpayer who bears the cost through higher power prices. The capital cost to build nuclear power is often 2 to 3 times the initial estimate, and the money is hidden in military or other such spending. In this section, we will look more closely at some detailed costs, as there are claims and counterclaims on the true costs.

4.12.1 Nuclear Operating Costs

The operating costs of nuclear plants are often hidden from the public and buried in seemingly unrelated costs. The actual operating costs are pretty low – the indirect costs of financing construction drive costs up.

- Fuel Efficiency: A single kilogram of uranium can produce 360 MWh of electricity. It's the equivalent of 120 tonnes of coal, with low carbon dioxide emissions.
- Fuel Cost: Lazard assumes a cost of $.85/MMBTU for fuel. Only 33% of the heat produced is converted into electrical energy. That 33% thermal efficiency is about the same as a coal plant. That means the nuclear fuel-only cost is about $9/MWh. The World Nuclear Association estimates this cost to be even lower, at less than $5/MWh.
- Safety Costs: While the generating costs are similar to thermal fuel,

the high level of safety demanded adds additional costs.

- Waste Costs: On-site storage is cheap, with used fuel stored in concrete barrels in swimming pool-sized ponds for 40 years. The long-term cost of storage is unknown, as detailed in a following section.
- Cleanup Cost: Costs of $350 million are the estimated costs of US plant clean-ups but this figure is hopelessly outdated.

4.12.2 New Nuclear Plants Costs

Nuclear plant construction is often characterised as exhibiting "negative learning." That is, instead of getting better at building plants over time, we're getting worse. Plants have become radically more expensive, even as technology has improved and our understanding of the underlying science has grown. The only country in which construction costs seem to have steadily decreased is South Korea. South Korea's outlier status has led some experts to speculate that the cost data (which the utility provides without an independent audit) may not be reliable.

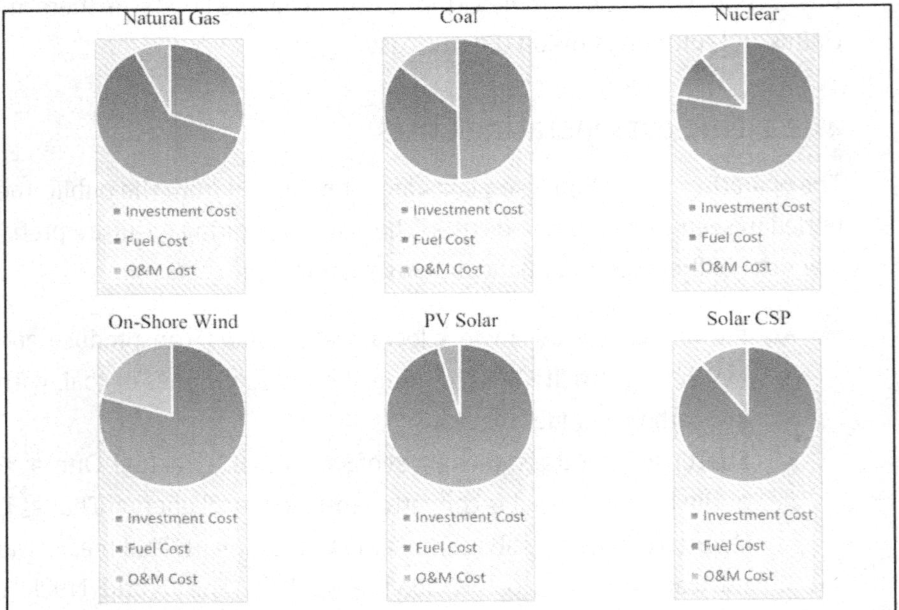

Figure 8. Percent contribution to LCOE of cost categories for coal, natural gas, nuclear, wind, PV solar, and solar CSP. Data from (Wittenstein and Rothwell 2015).

Figure 39 Contribution to LCOE from capital, operating and fuel costs

Figure 39 shows the breakdown of the costs of operating any power plant into three categories: fuel costs, operation and maintenance costs, and capital costs – the amortised cost from building the plant itself, including any financing costs. Gas, coal, nuclear, solar and wind have different cost breakdowns. For nuclear plants, 60-80% of their electricity cost comes from constructing the plant itself. [94] [95] For solar it is over 90% - because you construct the plant and have free electricity for the life of the solar. Gas and coal need less capital but they have ongoing fuel and operating costs.

There are 2 sources of costs of new plants. One is to use costs from authorities such as Lazard or CSIRO Gencost [81], [96], [97] and other analysis.

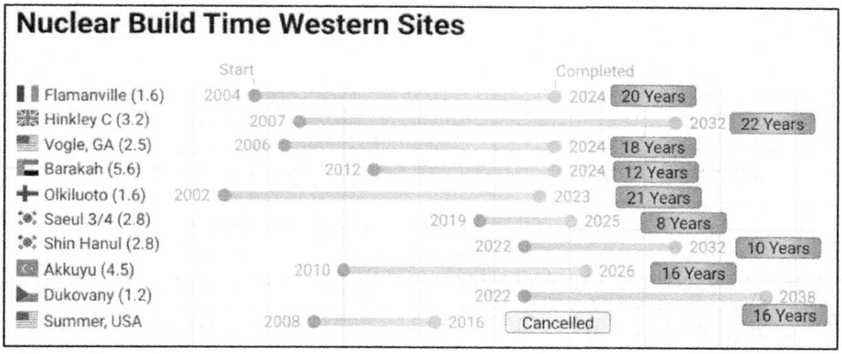

Figure 40 Nuclear build times for Western nuclear plants since 2000

Figure 40 shows the time for construction of western sites over the past 25 years. The South Korean are under 10 years but most take up to 20 years.

4.12.3 Nuclear Plants - Over Budget and Late

Why is there such a gap between what a plant will cost, and final price, and why are timeframes so wrong. Prof Bent Flyvbjerg in his book How Big Things get Done [98]_suggests it is the politics. Politicians under quote to get the project going – if people knew at the start they would never start. The biggest fat-tailed risks are nuclear storage and Olympics, where costs are often more than twice the original estimate.

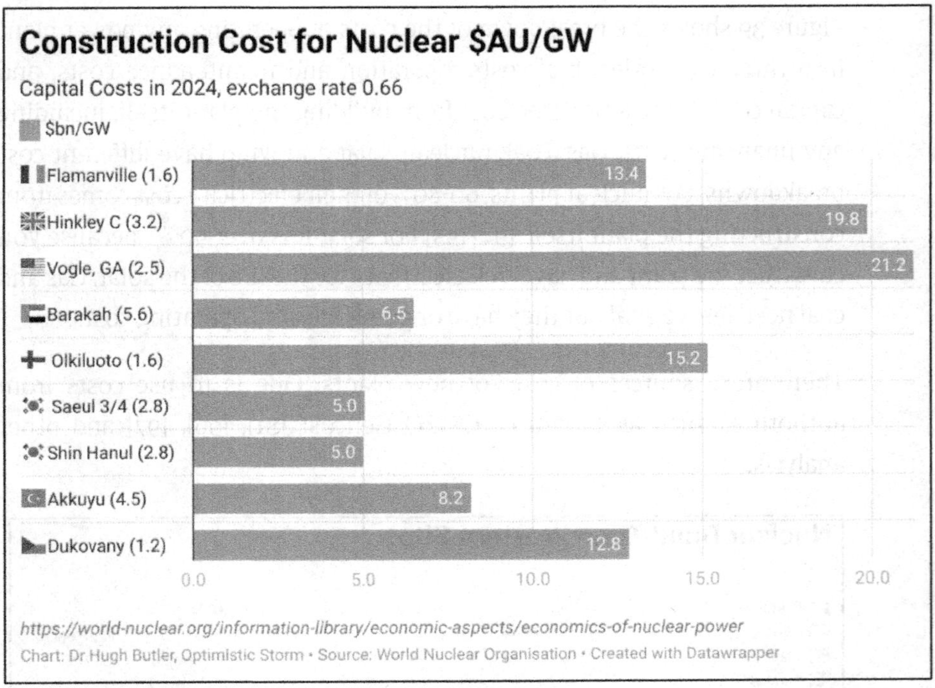

Figure 41 Cost of new build nuclear power plants for the last 9 Western plants (per GW)

Figure 41 takes those same Western nuclear plants and plots out their costs. The lowest is $4 billion in South Korea to a cost of more than $16 billion per GW for Vogle and Hinkley C (now much higher). In context, these numbers need to be compared with other expenditures. In Australia, the cost to build a dual lane freeway from Brisbane to Cairns, 1330 km would cost about $50 billion. Or the inland rail from Melbourne to Brisbane was on target to cost about $30 billion. When politicians talk about building nuclear, keep in mind these are eyewatering amounts of capital tied up for decades.

4.12.4 Nuclear First of a Kind Concept

We often see the concept of costs and the term First-of-a-kind (FOAK). Technologies almost always need early support as they are in direct competition with established technologies. This is an area where the government can, and has historically, contributed to technological innovation.

Early support for new technologies provides multiplying and sometimes compounding benefits to society. [99]

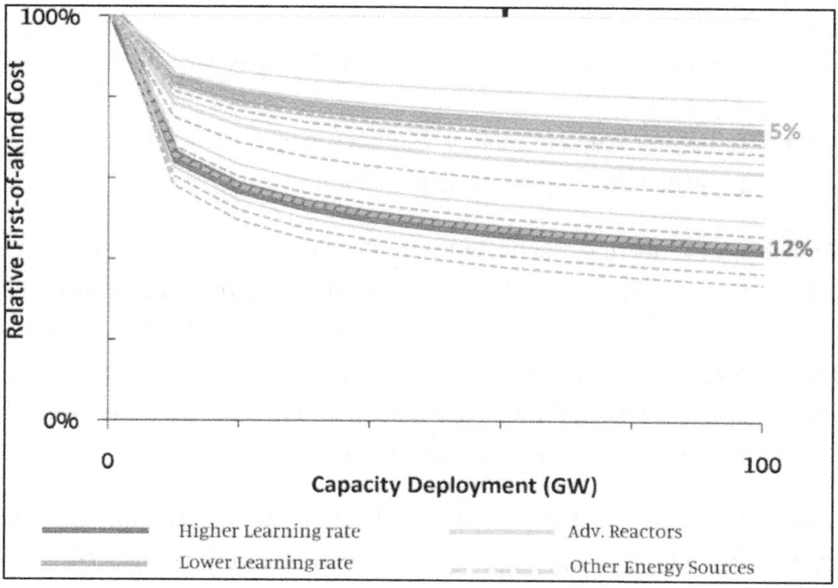

Figure 42 Are learning rates of nuclear likely to be at 5% or 12%

Figure 42 from the Breakthrough Institute shows the FOAK for advanced reactors, such as SMR plants. Building the first, the cost is 100%. As deployment increased from 1 up to about 8GW (i.e. 8 reactors), the costs should decline by 40%. This is the graph that nuclear advocates like to show. Real-world data show costs have actually increased. Examples include Hinkley C and Sizewell C. The learning rate is not 5% or even 12% but can be negative 10%. The more plants built, the more expensive they become.

4.12.5 French Nuclear

France is a good example of where technology and economics clash with politics, and from a nuclear-advocate perspective, it is an ideal country to compare with others, so we will dig a little deeper into the background and current status. All the same issues we have previously discussed are evident. Most advocates who point to France as a country that has made a success of nuclear power need to take a cold shower. France's decision has cost it dearly and will do so for decades to come.

France's history with nuclear power runs deep. The country operates 58 nuclear power plants with 45 built in the 1980's and they added some more in the 1990s:

- The building was largely built in response to the oil crisis in the 1970s.
- It has no natural fossil fuel resources in its territories.
- It needed them for its nuclear weapons.

France began shutting down its oldest nuclear plant in 2020 after 43 years of operation, the first in a series of closures the government has proposed, though the country won't altogether abandon its reliance on nuclear power. The closures are part of a broad energy strategy to rely more on renewable energy sources. That strategy would see French dependence on nuclear energy from supplying three-quarters of its electricity to about half by 2035.

Today, nuclear is the largest source of electricity with a generation of 380 TWh, or 71% of the country's total electricity production of 540 TWh. [100] It is the highest percentage in the world. Those numbers are in the good years. Since June of 2020, it has 56 operable reactors totalling 61,370 MW, one under construction (1630 MW), and 14 shut down or in decommissioning (5,549 MWe).

4.12.5.1 Status of French Nuclear Plants

Every year, an independent commission suggests the regulatory electricity price in France, and the answer is the same. The price has to increase to meet the cost of decommissioning the reactors. The answer is the same as the government recoils in horror and says, "Not on my watch." Between 1973 and 1998 France shut down seven reactors. Only one has been dismantled, and two are being decommissioned, one shut down in 1985, the other in 1998. That means the real-world decommissioning process exceeds 25 years, as the UK and USA have discovered. Successive governments have not been enthusiastic about speeding up the process because it means higher electricity prices.

4.12.5.2 Reliability

Often touted as reliable, the fleet is actually old (many 40 plus years) and the company has inadequate funds. In April, a report from The Times reported that the operator of France's electricity grid, Réseau de Transport d'Électricité (RTE), had asked customers to cut consumption as cold weather increases demand at a time when about half the country's nuclear reactors are offline. French nuclear reliability has been dismal. It was not generating for 96 days in 2019, 115 days in 2020, 103 days in 2021, and 152 days in 2022 due to droughts, insufficient water, or pipe repairs. We explore Frances' difficulty with cooling in the Cooling section.

4.12.5.3 Private Investment Failure in France

In most other countries, private investment has built fossil fuel plants or, more recently, renewable energy generation. Private investment runs a mile from nuclear because of uncertain returns and the risk to investors. In France, Electricité de France (EDF) is the major company for nuclear. Bloomberg reported [101] that the French Government was looking to nationalise debt-laden power company EDF. It holds an 84% stake in the $30 billion energy giant. In 2022, the Government offered $2.4 billion to support EDF, as the group saw a hit to profits due to outages at several of its nuclear plants and the impact of a government power price cap. From the most recent balance sheet, EDF had liabilities totalling €97 billion falling due within a year, and €201 billion due beyond that. It had cash of €43.4 billion and €40.9 billion in receivables for net liabilities of €214.7 billion. This debt towers over the €30.6 billion company. The French Govt is locked into supporting EDF. In May 2022, they say, "in accordance with its commitment, the French State subscribed for an amount of approximately €2.7 billion, representing approximately 83.88% of the Rights Issue and holds approximately 83.88% of the Company's share capital following the completion of the Rights Issue." It was a sleight of hand for the French Government to hand them another subsidy.

4.12.5.4 New Nuclear In France

As with many political announcements, things don't always go to plan. For example, President Emmanuel Macron announced plans to build six new reactors, and to consider building a further eight. The President highlighted the need to increase electricity supply by "up to 60%" as the country seeks to reduce oil and gas consumption over the next 30 years.

> *"Key to producing this electricity in the most carbon-free, safest and*

most sovereign way is precisely to have a plural strategy... to develop both renewable and nuclear energies. We have no other choice but to bet on these two pillars at the same time. It is the most relevant choice from an ecological point of view and the most expedient from an economic point of view and finally the least costly from a financial point of view."

President Macron 2022.

- A year later, in January 2023 the French Senate approved a draft bill – 239 votes in favour and 16 against – to accelerate procedures relating to the construction on new nuclear facilities and the operation of existing facilities. The bill included:
- Removal of the objective to reduce the nuclear share of France's electricity production to 50% by 2035, instead maintaining it at more than 50% by 2050.
- Constructing new EPRs and SMRs by 2050.
- Exemption from certain urban planning permissions for the construction of new reactors.
- Option to use an immediate possession procedure to obtain land on which to build new reactors.
- In March 2023 France's Parliament formally approved the government's nuclear investment plan – by 402 votes in favor and 130 against – which considers the €52 billion construction of six new EPR-2 PWRs at three sites. President Macron plans to begin construction of the first reactor before the end of his current term in May 2027. Look at that number again. Hinkley cost more than that number, and the same builder reckons prices will come down 3 times?

It is not all beer and skittles. France's national audit body warned that the task of building six new nuclear reactors in France — similar in scale to a previous Prime Minister in Australia's vague plan for seven reactors of various kinds around Australia —was not currently achievable. [102] Initial estimates of €51.7 billion was revised up 20% to €67.4 billion ($AU112 billion) in 2023. While the plan was to start in 2027, it was unclear how the project would be built, as supply chains were not in place. The proposed model is a new simplified EPR model, known as EPR2. EDF lost a bid for 2

Czech units, as well as the 2009 UAE plant, to the Korean KHNP design. Furthermore, no one knows how it will be financed, with little commercial interest prompting the French government to consider an interest-free loan to EDF.

- By July 2024, EdF, facing potential bankruptcy due to delays and massive cost overruns at Flamanville and Hinkley's latest generation large-scale nuclear plants, scrapped plans for a new design for SMRs.
- A month later, EDF cancelled the planned nuclear power plant and turned off 6 nuclear reactors.
- Meanwhile, Macron's plan stated 14 of the country's nuclear reactors will shut down by 2035 and 4-6 of those by 2030, without any plans for anything else.

EDF is currently building Hinkley and planning to develop Sizewell in the UK with all those costly overruns and risks. They experienced their "Hinkley" in France with major cost and delays.

- Flamanville in Normanby was started in 2007.
- Initial completion estimate was 2012. EDF turned it on in December 2025 and expects to be in full production in mid-2025.
- Initial estimate of cost was €3.3 billion and came in at about €13.2 billion or €19 billion in 2015 terms.

4.12.5.5 France's Performance Numbers

France also has cost, waste, and cooling challenges.

- Waste - by 2013, about 370,000 tonnes of spent fuel had been produced.
- Less than 120,000 tonnes had been processed.
- That leaves 250,000 tonnes with more being produced every year;
- Global processing capacity is only 5,000 tonnes per year.
- France had over 1,000 incidents involving reactors in 2019 alone.

4.12.5.6 France – Part of European Grid

Fortunately, France is connected to multiple countries by extensive electricity grid networks with plans to increase interconnectivity. Spain to the south has an abundant solar surplus, and Scandinavian countries have hydro. In spite of having nuclear (low emissions), it actually has one of the highest emissions in Europe due to its gas power stations.

4.12.5.7 France – Low Emissions

France has had lower emissions than other European countries, UK and Australia over the period 1970s to 2020's but has not made much progress over the past decade. In contrast, the UK, with the construction of offshore wind farms and the closure of the last coal-fired power plant, is still reducing emissions rapidly. Australia emits 14 tonnes CO_2 per person, Germany 6.67, UK 4.53 and France 3.97,

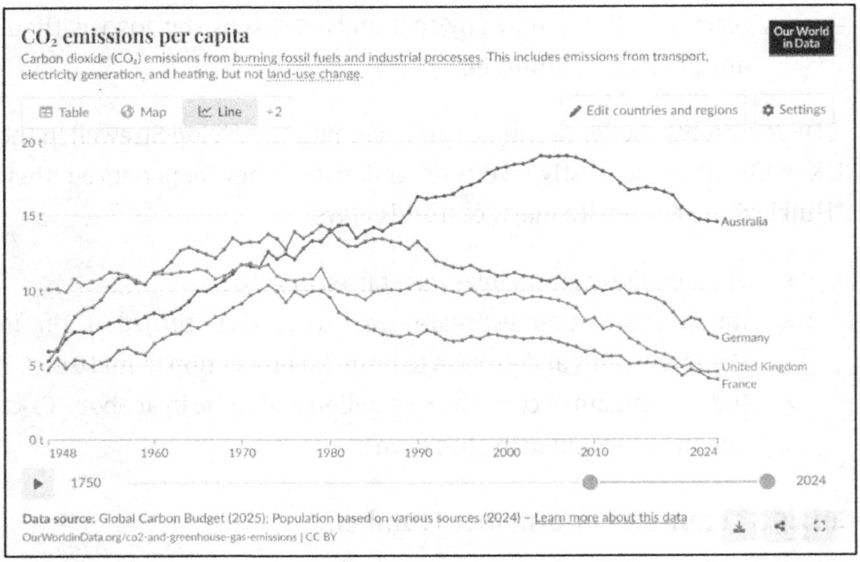

Figure 43 CO2 emissions per capita France, UK, Germany, Australia

Figure 43 shows that the fall in per capita emissions in France have only slowly reduced. As France's population and energy consumption increased, it turned to gas and oil to make up for the constraints of nuclear power and has not made much investment in either wind or solar.

4.12.6 USA Subsidies

The USA provides large subsidies. From subsidising enriched fuel to keeping plants open. Providing those funds still does not solve the waste problem. Here is a short list of subsidies in the US:

- Grant to PG&E over $1.1 billion to keep Diablo Canyon open for a further 5 years after 2024/2025. [103] Diablo provides about 8.65% of California's electricity.
- $4.3 billion to build production facilities for enriched uranium to wean off Russian enriched uranium. [104]
- $370 billion in the IRA with a $15/MWh tax credit for existing reactors supporting $42MWh total compensation. This program is in doubt.
- $30 billion in production tax credits.
- $6 billion to support at-risk nuclear power plants. The bailout is seen as misguided by many. [105]
- $3.5 billion for advanced nuclear power.
- $41 billion for Vogtle for 3GW, $12 billion of which is loan guarantees. The price of electricity is $112/MWh and $189/MWh. The Georgia Govt allowed $2.1 billion of the extra build costs to be billed directly to customers, increasing their power bills by $3.78 per month.
- $1.8 billion for 3 Mile Island restart.
- $1.52 billion loan for the restart of Palisades 800MW. Additionally, $1.3 billion in grants to rural electric cooperatives near the plant to purchase the more expensive power.
- $1.1 billion for 10 years of Pacific Gas & Electric's Diablo Canyon, which was due to shut in 2025. California to cancel $400 m loan to Diablo.
- $494 billion for cleanup of nuclear repositories.
- Department of Defence and Department of Energy plans to operate, sustain, and modernise current nuclear forces and purchase new forces will cost $946 billion over the 2025–2034 period, or an average of about $95 billion a year. That is up 25% from 2023.

Subsidies are often hidden. The Department of Energy Awards grants have surpassed $1 billion in funding since 2009 to support nuclear energy research and development at U.S. colleges. In their 2025 Congress Appropriations, [106] $1.3 billion for nuclear research. $12 billion for nuclear waste. $80 billion on facilities such as Oak Ridge, Los Alamos, and Savannah River. It is variously estimated that over 60% of the $161 billion budget for the DOE is focused on nuclear - research, regulation, weapons, and industry support. Under the Trump Administration it cut clean energy research.

4.12.7 Finland Nuclear

Finland has 5 operating nuclear reactors and 1 planned, with nuclear contributing about 60% of electricity. Finland's prime reason is to replace coal. The mix of electricity in Finland in 2022 was:

- Nuclear 23.3 TWh (34%)
- Hydro 15.9 TWh (23%)
- Biofuels & waste 11.9 TWh (17%)
- Wind 7.9 TWh (11%)
- Coal 5.5 TWh (8%)
- Gas 3.7 TWh (5%)
- Oil 0.3 TWh
- Solar 0.3 TWh.

Construction is typical of nuclear plants beset with cost overruns and delays. For example, the construction of Olkiluoto 3 started in 2005. The completion date was blown out from 2009 to late 2021. 12 years late. However, after producing electricity for about a week, it had to shut down for another 3 months. Then, after ramping up to about 30% of the capacity, it encountered another problem, delaying it for another 5 months.

- In June 2022, 17 years later, Olkiluoto 3 provides exactly 0 MW to the Finnish grid
- In Finland, the wholesale cost of power has been about €50 /MWh for the past decade, but increased over past 2 years to over $100 /MWhr.
- Spikes have been as high as €400, and at peak times to €10,000 due to with poor rainfall, import of power and gas from Russia, and strong demand.
- Waste disposal cost exceeds €2.3b. Cost is direct from general funding.

Finland receives nearly as much solar radiation annually as northern Germany, where solar is widely deployed. Long summer days allow for high production, and snow reflection in early spring boosts efficiency. Cool temperatures also improve panel performance, as panels operate more efficiently in cooler climates than in extreme heat. By early 2026, just 4 years after completion of Olkiluoto 3, Finland had a combined capacity of domestic wind power and utility-scale solar power of 10 GW. This is three times Finland's total hydropower capacity and corresponds to roughly 6–10 large nuclear reactor units. The question will be be if Finland continues to build out more nuclear plants or will change to what the rest of the world is doing.

4.12.8 UAE Nuclear

UAE have built four NPPs that may end up producing at $80/MWh ($0.08kWh). Unfortunately, the UAE is also building large-scale PV that will be producing at $13 MWh ($0.013/kWh). And since they're still burning gas for most of their power, every kWh from solar goes straight to reducing the overall cost and CO_2 emission, five times cheaper than the NPPs will.

4.12.9 Canada Nuclear

Many say that Ontario nuclear in Canada is an example of cheap emissions-free power. That is the headline but the reality is quite different. Ontario highly regulates its electricity prices. Nuclear power generators don't need to recover their costs via a competitive wholesale market. Consumers pay a specific regulated levy to Ontario Power Generation for it to recover the costs associated with its nuclear power plants. The decision and order EB-2020-0290 sets out the total amount Ontario Power Generation can recover from consumers each year from 2022 to 2026. [107] The total nuclear payments range from $70 to $85/MWh over 2022-2026. These payments for nuclear power plants built several decades ago. Previously, the state wrote off most of the construction cost because it was unsustainable debt. In reality, Ontario's budget for reducing electricity prices exceeds spending on long-term care and is in excess of $6 billion per year, [108] up 4 times from 2018. The subsidies knock $50 off the average monthly residential electricity bill. Without the subsidies, the typical household would pay $171 a month (including tax) for electricity. Instead, the average hydro customer's monthly bill is $121. Ontario (2019) zero-carbon sources exceeds 92% with the following contribution to emissions

- Nuclear=59%
- Hydro=24%
- Wind=8%
- Solar=1%

Much of the subsidy goes to businesses, including smelters. How much is from previous green energy projects and how much from subsidising nuclear is murky.

The history is that Ontario embarked on a massive building spree between the 1960s and the 1990s. As Winfield says in 2024, "Private capital won't touch it and the cost curve just keeps getting steeper – unlike renewables and batteries". The government of Ontario is planning to commit the province to a $100 billion nuclear construction and reconstruction program with four small modular reactors (SMRs) from GE's Vernova Hitachi Nuclear Energy's BWRX-300 SMR. They have no license yet. There are no existing SMR approvals. Existing cost estimates are $14 billion or more [109]. That is for a total generation capacity of 1.2 GW, the equivalent of a single large coal-fired power station, and only about 5% of the outgoing coal generation capacity (21 GW), which Australia needs to replace over the next decade or so. The timeline for the first unit is to complete construction by the end of the decade and be in service in 2030. The FOAK cost is estimated at $5.6 billion, and they have assumed a 70% cost reduction by the time the third reactor gets built. The remaining three units will be completed in the mid-2030s, and indications are that the cost of electricity will be more than $140/MWh. The initial SMR cost is comparable to that of the Vogtle nuclear plant, the most expensive power plant ever constructed.

4.12.10 China Nuclear

One would expect China to be different, and it is helpful to understand the history of China. Their story is that for the first decade, they used French technology. German technology followed, and now they have done it themselves. In 2011, China declared it would make nuclear power the foundation of its electricity generation for the next 10 to 20 years and add over 300 GW of capacity. (about 2,300 TWh). That's a 7 times increase over the existing electricity supply. We can look at the past, but it is more difficult to project the future when plans change every 5 years with their national 5 year plans.

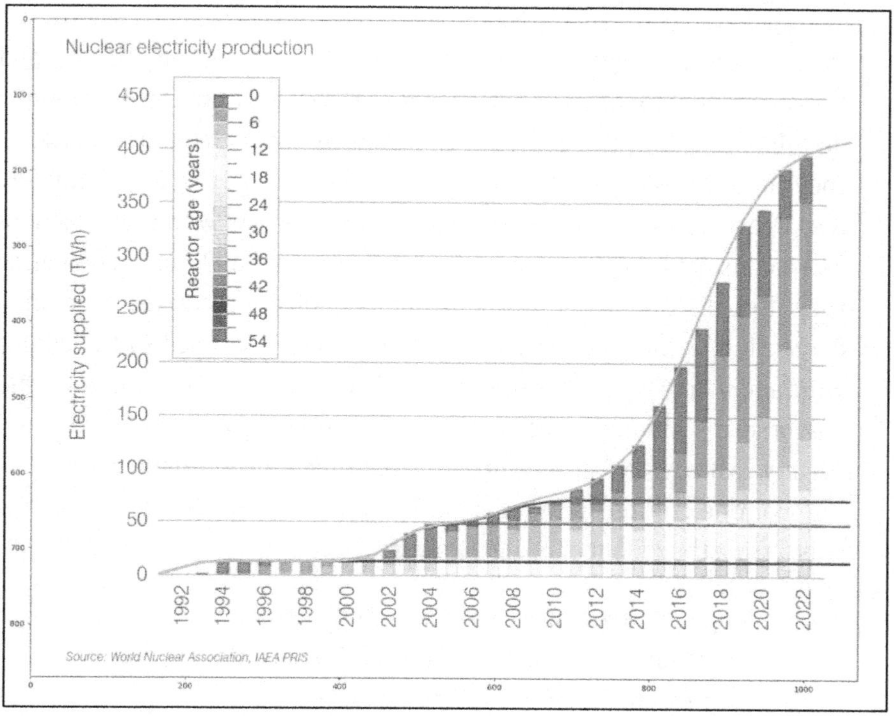

Figure 44 China electricity supplied by nuclear TWh from 1992 to 2022

Figure 44 shows electricity supplied (TWh) from 1992 through to 2022. There was a rise from 2002 to 2006 and a tripling from around 2006 to 2020. They reviewed their plans after the Fukushima disaster and stated that they would achieve 15% of China's total electricity generation by 2035, 20-25% by 2050, and 45% by the second half of the century. By 2023, they were 5 years behind in nuclear, but five years ahead in renewables from solar and wind. Nuclear power accounted for 4.9% of the total electricity mix in 2019, and was expected to grow to about 10%. By 2024, China's nuclear power generation reached 445 TWh or 4.72% of its total electricity generation with 57 nuclear power units in operation with a combined capacity of close to 60 GW. Additionally, there were 27 reactors under construction with a total power of 32 GW. In reality, nuclear supplied a smaller percentage of electricity over that 5-year planning cycle. It grew, but not in percentage terms. China then said that it plans to invest $440 billion in 150 new plants by 2035. The 200-gigawatt plan would generate enough power for over 260 million people. However, it isn't easy to get reliable, quantifiable information from the PRC on how those plans are progressing.

Recent reports suggest they have reduced the focus on nuclear and transferred that to solar and batteries as they see that as the fastest route to decarbonisation and to supply the annual increase in demand. [110] In 2023, they aimed to reduce nuclear power to 10% by 2035, adding approximately 6-8 large reactors per year. Their solar deployment is now the equivalent of 40 reactors per year. The data suggest their nuclear build-out is slowing, as they, too, discover that the cost of nuclear plants increases over time. (increasing learning curves). The costs have not decreased to the extent wind and solar have. Cost is the primary driver. Between 2011 and 2022, the cost of solar PV modules declined by 85%, wind energy costs by 60-70%, and battery costs by 90%. It now appears that China will drop from being the number one emitter to number six by 2030. It is unclear why China is reassessing the rollout of nuclear power.

- Rapid deployment of wind, solar, and batteries.
- Substantial investment in large-scale transmission.
- Refurbishment of coal plants for backup.
- Deployment of batteries.
- Decisions by provinces versus central party.
- Price.

Or it could be that China generated 18% of its electricity from solar and wind in 2024, twice as much as in 2020 (9%). When you look at the most recent numbers in 2025 the difference is stark.

- New nuclear = 2 GW - 60 GW total capacity
- New coal = 95 GW – 1,500 GW total capacity
- New wind = 119 GW – 660 GW total capacity
- New solar = 250 GW - 1,240 GW total capacity
- Storage - batteries = 76 GW has overtaken pumped hydro of 60 GW

Nuclear is a laggard even to coal, and a tiny fraction of their wind and solar deployments. and is adding less than 10 TWh (about 1GW). China faces a rapid increase in electricity demand, and renewables are making up that new electricity, not nuclear.

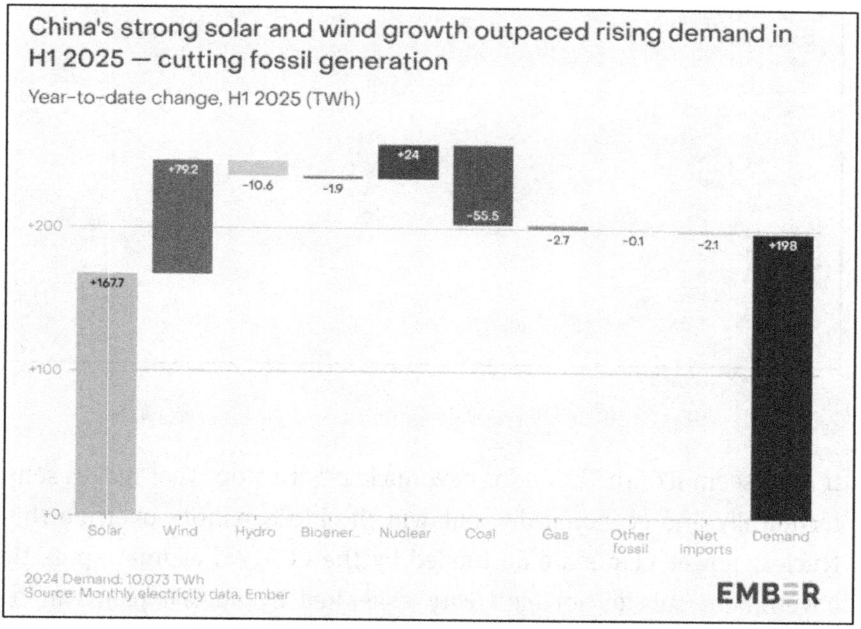

Figure 45 China new electricity demand 2024 & supply change

Figure 45 shows that electricity demand increased by over 200 TWh. In context, that's about the size of total Australian east coast electricity market (211 TWh). Solar met 167 TWh and Wind 79 TWh. Nuclear added 24 TWh. The consequence is coal fell by 55 TWh. That is good news for the world, as it appears coal has plateaued in China, despite the fact that they are continuing to build or replace coal power stations. The capacity factor is low, and the main reason for building them is the ability to provide electricity when the sun does not shine and the wind is not blowing. But if we look at the rise of grid-scale battery deployment (over 65 GW), this will provide the grid backup and push out coal even faster.

Figure 46 below demonstrates that China is also suffering from longer approval times for nuclear, with a typical of six years. They are not a democracy so many of the constraints in the West do not apply. They also have a strategy of sovereignty and energy self-sufficiency, which is why you don't see a large build-out of gas infrastructure, since they don't have large reserves within China.

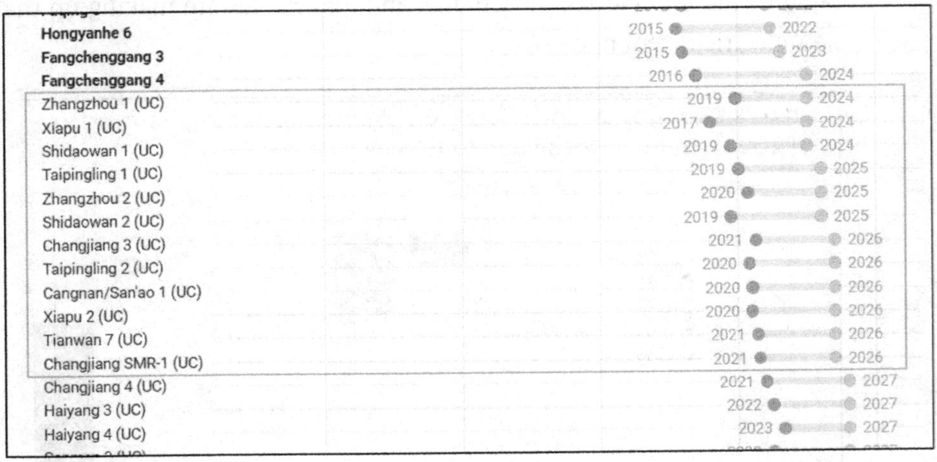

Figure 46 Timeline Chinese nuclear power plants from approval to construction

It does seem like the trend for new nuclear is to stop. That makes sense technically and economically, but will the PRC's politics overrule this? Nuclear power plants are all funded by the CCP. An estimate puts the government subsidy for electricity generated by nuclear plants at 0.4 yuan/kWh ($60/MWhr). This is similar to nuclear electricity subsidies in the west and about the cost of wholesale electricity from wind solar and batteries. What are their political drivers to build nuclear at twice the cost? Or will they also come to understand that wind, solar, hydro and batteries are a robust solution to electricity supply.

4.13 Risk and Insurance

Nowhere in the world will insurance companies take on the risk of nuclear alone. Every country with a nuclear generation fleet has enacted legislation that caps private liability at some level and shifts any liability above that level to taxpayers. In the US, at present, that's $13 billion for the entire sector, around $131 million per reactor. These sound like big numbers, but as was pointed out, Fukushima's total liability is still moving to the trillion range. The number of countries willing to take on that liability is shrinking globally, not growing. The operators of nuclear know this. It is part of the subsidies as above. As the CEO of Duke Energy [111] said in a quarterly shareholder report:

" we operate the largest regulated fleet of nuclear plants in the country. We have - we think nuclear has a place to play and a lot of promise in the future. But before we go down that path, we're going to have to have some things figured out. We're going to have to resolve the first-of-a-kind risk, design, supply chain, and workforce issues for SMRs and even bigger reactors, and determine how we're going to handle that. We're also going to have to have overrun protection from the federal government or others to be able to protect our customers and our investors from any overruns on these projects. And then lastly, we're going to have to have a means to make sure that we're protecting the balance sheet as we're building these facilities. So until we get those items resolved, we're still looking at solar, gas and upgrading and getting everything that we can out of our current assets."

The real cost of past accidents will be expanded in a later section.

4.14 Cooling Challenge

Most nuclear plants are built near the sea or rivers because they require large amounts of cooling water. A nuclear power plant works much the same as a coal-fired station, using an energy source to boil water and produce steam that spins a turbine. During this process, water is used to cool the reactor, generate electricity, and condense the steam produced back into water.

Plant and Cooling System Type	Water Withdrawal (L/day)	Typical Water Consumption (L/day)
Coal, once-through cooling (700 MWe)	1.5 billion to 3.8 billion	7.56 million to 24.4 million
Coal, closed-loop cooling (700 MWe)	38.6 million to 92.4 million	37 million to 84 million
Nuclear, once-through cooling (700 MWe)	1.9 billion to 4.6 billion	7.56 million to 30.2 million
Nuclear, closed-loop cooling (700 MWe)	60.5 million to 84 million	44.5 million to 64.7 million

Figure 47 Water usage for coal plants and nuclear plants using once-through or closed-loop cooling

Figure 47 shows a nuclear reactor uses more water for cooling than a coal plant. Only about 30% of the heat energy actually generates electricity. Without enough water supply, known as a heat sink, a reactor must be shut down to prevent overheating, which in extreme cases can cause a nuclear meltdown. The rest of the heat energy has to be cooled or dispersed. Water withdrawal and consumption for coal or nuclear using either once-through cooling or closed loop cooling shows that typical water consumption is over 50 million litres per day. [112] The once-through cooling method poses a greater risk to river ecosystems if water temperatures become too high. This is the problem with French nuclear. These Australian numbers align with the Nuclear Energy Institute estimates that, per megawatt-hour, a nuclear power reactor consumes between 1,514 and 2,725 litres of water. Once-through water use is 25,000 to 60,000 litres per MWh, and recirculating consumption is 600 – 800 litres. The hot water released (not the steam) can be a serious problem in downstream waterways. Nuclear power plants already have restrictions due to limited water, yet building them close to oceans with high risk of sea level rises seems bizarre. France lost much of its capacity due to droughts and high temperatures in the rivers used to cool. Building plants next to oceans has its downfalls. China, with its nuclear program, is building its reactors close to the sea, except for its Thorium reactor built in the Gobi desert. The thorium reactor relies on molten salt and requires less water to operate. Plants built near the ocean face a different set of risks. Ask Fukushima residents. That is without any sea level rises from Climate Change.

4.15 Waste

The nuclear waste problem is growing. Some say calling irradiated nuclear fuel "waste" is wrong given that more than 90% of the fissionable energy remains unused. There is no effective method to refine the waste. Arguments come in 2 flavours.

- Not recognised: The scale of the problem is not recognised.
- Recognised, but trivialised: some nuclear advocates trivialise the biggest and increasing problem in an attempt to deceive those less informed.

Building a nuclear reactor without first having a waste disposal plan is like building a house without a toilet. We may hope there's a plan for the toilet so we don't have to fight the cat for access to the kitty litter. Essentially, there are five issues.

- Radiation illness and death.
- Longevity of the radioactivity for more than 100,000 years.
- Quantity - over hundreds of thousands of tonnes.
- Stability of storage site.
- Cost – billions – paid for by the taxpayer.

4.15.1 Dangers - Radiation and Exposure

All waste issues result from the danger posed by radiation and how to protect ourselves from it. Marie Curie's research into radioactivity, while groundbreaking, also exposed her to significant levels of radiation, leading to her death from aplastic anemia, a condition now understood to be caused by radiation exposure. Initially, the dangers of radiation were not fully understood, and both she and Pierre Curie worked with radioactive materials without adequate protection. Their experiences, along with those of others in the early days of radiation research, helped pave the way for understanding and managing radiation sickness. With the atomic bombs in Japan, the United States military estimated that around 70,000 people died at Hiroshima, though later independent estimates argued that the actual number was 140,000 dead. [113] At Nagasaki, between 40,000 and 80,000 died. In both cases, the majority of the deaths occurred on the day of the bombing itself, with nearly all of them taking place by the end of 1945. There was no real tracking of subsequent deaths from radiation. Medical facilities were limited and we did not have the techniques and ability to track subsequent deaths. Were there deaths over the next decades?Absolutely! What you don't measure you cannot report on and for many reasons most wanted to forget those killings. We do know military personnel exposed to above-ground test explosions through the mid-1950s did suffer radiation and cancers, and those in the development of nuclear reactors or in subsequent nuclear accidents are the prime source of our knowledge of radiation toxicity. We now know that:

- Nuclear bombs produce mostly short-lived isotopes and radiation.
- Nuclear power plants produce many more isotopes and those isotopes are much longer-lived.

4.15.2 Half Life of Nucleotides

The concept of half-life is really important in discussions of waste and accidents. Do we need to store things for 5 weeks, 5 years, 50 years, 500 years or 5 million years? Radioactive isotopes eventually decay, or disintegrate, to harmless materials. Some isotopes decay in hours or even minutes, but others decay very slowly. A half-life is the period after which half of the radioactivity has decayed. If a half-life is 20,000 years, in 100,000 years, that will have decayed five times, or to 1/64th of the original amount. Another 20,000-year half-life would take it to 1/128th of the amount. Here are some examples.

- Half life of ^{131}Iodine is 8 days.
- Half-lives of some elements even within one element vary hugely. For example, $_{55}$Caesium has 41 known isotopes, ranging in mass number from 112 to 152. Only one isotope, ^{133}Cs, is stable.
- ^{135}Cs with a half-life of 1.33 million years
- ^{137}Cs with a half-life of 30.04 years
- ^{134}Cs with a half-life of two years.
- All other isotopes have half-lives less than 2 weeks and most under an hour. [114]
- Half-life of ^{245}U and ^{239}plutonium (Pu) are 24,000 years
- ^{90}Strontium half-life is 30 years.
- Some think it is just uranium reactivity that in a reactor, but the radiation from the uranium reactions also creates a huge range of elements in the waste, the steel and the infrastructure, some with half-lives of over millions of years.

Very few countries have found ways to store waste that is environmentally acceptable and acceptable to the population. Primarily, waste is stored on site. Iceland believes it has appropriately determined a site, the process and economics for the next 100,000 years. However, there is no certainty that the site will survive 100,000 years. Spent fuel rods from Olkiluoto's nuclear power plants cool off for several decades in interim storage pools before final burial at Onkalo. [115]

4.15.3 Radiation Dose

The sections on Chornobyl and Fukushima will explore this in more detail but firstly some understanding of radioactivity and exposure. The media is often confusing and mixes up SI units and common units. Radiation is invisible and we cannot use any of our sensing facilities to ground ourselves. Firstly, there is the radioactivity, and the type. Then we have the absorbed dose or exposoure. A dose equivalent depends on both.

	SI Units*	Common Units
Radioactivity	becquerel (Bq)	curie (Ci)
Absorbed Dose	gray (Gy)	rad
Dose Equivalent	sievert (Sv)	rem
Exposure	coulomb/kilogram (C/kg)	roentgen (R)

Figure 48 Radioactivity doses in SI units and common units

Figure 48 shows the various units used for each of these parameters. The annual limit for radiation exposure for a member of the public is 1 millisievert (mSv) per annum (1,000 µSv) with the average about 2.4mSv. Ambient levels of background radiation in Australia will vary between 0.05 µSv/hr to around 0.2 µSv/hr. Some places can be even higher than this if they have certain minerals around. The guidelines for a radiation worker is about 20 mSv per year – 20 times higher than background.

- A single chest x-ray exposes the patient to about 0.1 mSv. This is about the same amount of radiation that people are exposed to naturally over about 10 days.
- A mammogram exposes a woman to 0.4 mSv, or about the amount a person would expect to get from natural background exposure over 7 weeks. That is why breast screening is done every two years – one does not want to cause breast cancer by over-zealous exposure.

- A seven-hour aeroplane trip exposes passengers to 3μSv or 0.002 mSv.

At Chornobyl, the average effective dose for 530,000 recovery operation workers was 120 mSv, with some receiving much higher doses. The 115,000 evacuees received an average dose of 30 mSv – about 30 years' worth at once. [116] Note that there were over 100 radioactive elements. Most are short-lived and decayed (reduced in radioactivity) very quickly. Iodine, strontium and caesium were the most dangerous of the elements released and have half-lives of 8 days, 29 years, and 30 years, respectively. ^{90}St and ^{137}Cs are, therefore, still present in the area to this day.[117]

From an ethical perspective, we cannot subject people to harmful radiation in scientific trials; therefore, we only have observational studies to understand long-term illnesses and mortality associated with radiation exposure. For a long time, it was believed that there was a minimum threshold of exposure, but that now seems to be false. Risk is a function of dose times and exposure.

There is no safe level.

4.15.4 Scale of the Nuclear Waste Problem

You may have seen a nuclear advocate holding up a Coke can and stating that the size of the problem is that. It fails to recognise the scale. Many Coke cans form a mountain and can generally be collected, melted down, and reused. An aluminium Coke can be returned to a can recycler for $0.10. Not so a can of used fuel rod. These issues are real with no solution.

- Quantity: There are over 267,000 tonnes, at about 2kg per person per year. About 10% or 22,000 tonnes of that is hazardous.
- Global: 14 countries across the world.
- Storage Stability: The life of a storage site is beyond reasonable confidence in geological science.
- Cost: Storing it all, according to GE-Hitachi, is more than $100 billion, (discounting China, Russia, and India).

4.15.5 USA Waste

In the U.S. alone, commercial nuclear power plants have produced more than 88,000 metric tons of spent nuclear fuel. Additionally, they have substantial volumes of intermediate and low-level radioactive waste.

- The General Accounting Office (GAO) estimates the total cleanup cost for the radioactive contamination incurred by developing and producing nuclear at a staggering $494 billion [118].
- Cleaning up Hanford has already cost taxpayers $170 billion over 30 years, but government auditors say the most challenging parts of the clean-up work are yet to be done over the next 50 years. It could be a legacy of $300 billion to $640 billion over 70 years to deal with 4 million curies. [119]
- The Savannah River site in Aiken, South Carolina, has 3 million curies in about 130 million litres of water. Note: Given the annual radiation limit for humans is 2,000-9,000 microcuries, those 3 million curies would exceed the annual doses for every man, woman, and child in the USA for a decade.
- Spent fuel, the most highly radioactive waste, will have to be isolated in deep-mined geologic repositories for hundreds of thousands of years.
- The USA has no program now to develop a geologic repository. It spent decades and billions of dollars since 1987 on the Yucca Mountain site in Nevada and now given up.
- Spent nuclear fuel is currently stored in pools or in dry casks at reactor sites.
- Spent fuel accumulates at a rate of about 2,000 metric tonnes per year.
- Decommissioning costs are about 9-15% of the initial capital cost of a nuclear power plant, or between $300 million and $400 million per plant [120]. With modern plants costing $10 billion per GW, that is a billion per plant. The decommission cost is a billion-dollar, 100-year venture. That's been shown empirically by reactors under decommissioning in several countries. The US holdback accounts for roughly a third of the "estimated" costs, so the taxpayer will be on the hook for the rest. Those "estimated costs" are fantasy

compared to real-world costs, and hopelessly inadequate. For example, the US Government collects about 0.1c to 0.2c per kWh during the life of a nuclear plant. The official line is that there is about a $9 billion shortfall in the USA, which the Nuclear Regulatory Authority is confident it will collect from the 100 operating plants. Others put the shortfall at over $70 billion. Indian Point in the USA alone is over $2.3 billion. A technique called Safe Enclosure ('Safstor') or deferred dismantling postpones final removal of controls for 40 to 60 years. Alternatively, Entomb entombs anything reactive in concrete. While a plant closes, the legacy lives on. For hundreds of millions of years.

Choosing a site is a fraught political process. For example, the State of Texas voted to stop the creation of a further waste site. A month later, the Federal Govt enacted a law to store nuclear power waste in Texas – similar to existing systems.

4.15.6 The £131 Billion UK Waste Problem

The £131 billion (estimates range from 99 to 232 billion) problem of how to safely dispose of vast volumes of radioactive waste created by the last British atomic energy program remains unsolved. [121] Nowhere in the economic assessment of nuclear power does this small economic issue feature. The ugly legacy of nuclear is laid bare in the UK "nuclear provisioning" for decommissioning, disposing of waste and remediation as touched on previously. [122] The ugly numbers from UK are:

- The Nuclear Decommissioning Authority owns 17 sites.
- Costs currently £3 billion annually.
- The legacy of 1940 to 1970s will go on for 120 years.
- One site, Sellafield, is 75% of the cost.
- Over 60% is met by the Government and taxpayers.
- The 2019 forecast is £124 billion over the next 120 years but could be between £99 and £232 billion or a discounted provision of £131 billion.
- In 2022, the cost of a proposed new UK underground nuclear waste facility doubled to £53 billion.

- UK currently stores its 700,000 m^3 of toxic waste, with over 133,000 m^3 of radioactive waste above ground. The quantity is projected to swell to more than 4 million m^3 in future.
- A larger volume of waste and 'more realistic' scope of costs has resulted in a bigger price tag for building and operating a long-term underground storage facility for radioactive waste.

4.15.7 No Solution for Australian Waste

Australia has been trying to establish a site – even for medical nuclear waste and recent efforts are now in court. Each year, Australia produces about 45m^3 of radioactive waste arising from the manufacture of the isotopes – about 40 m^3 low-level waste (LLW) and 5 m^3 intermediate waste. Australia signed up to purchasing some nuclear-powered submarines, but has no practical solution to the disposal in 30 or 40 years. That is no different to every other country embarking on additional nuclear without any practical waste solution.

4.15.8 France Waste Pas Mon Problème

As discussed in the previous section on France's nuclear program, it has about 59 nuclear reactors, producing more nuclear waste per capita than any other country – about 2 kilograms of radioactive waste per person each year.

- Following recycling operations, 96% of spent nuclear fuel (95% + 1% Plutonium) can be reused to manufacture new fuel, which will then supply more electricity in turn.
- High-level radioactive waste (4%) is vitrified, then conditioned in stainless steel canisters and stored at the La Hague site, pending disposal. No solution for the steel canisters has been found.

4.15.9 Fast Breeder Solution to Waste?

Nuclear waste could generate up to millennia of zero-carbon electricity for Europe if it were recycled and repurposed as fuel for advanced nuclear reactors, according to the environmental campaign group RePlanet and reported in Energy Monitor. [123] Spent nuclear fuel, the most radioactive part of nuclear waste, could be reused in advanced nuclear plants known as 'fast reactors' to create between 600 and 1,000 years of carbon-free electricity for the EU. A fast reactor is one in which the fission chain reaction is sustained by fast neutrons (carrying energies above 1 MeV or greater), as opposed to slow thermal neutrons used in thermal-neutron reactors. Instead of being buried in deep geological repositories, the spent fuel could be maintained in an accessible form and used to power the clean energy transition, argues the 'What a Waste' campaign. If all that money spent trying to find deep repositories was used instead to advance fast breeders, would be be in a different place? Could a burden be turned into a valuable part of a legitimate circular economic activity? Not likely

4.15.10 Novel Destruction Processes of Waste

Nobel Laureate physicist Gérard Mourou mentioned in his wide-ranging Nobel acceptance speech that lasers could cut the lifespan of nuclear waste using fast pulse lasers.

> *"Take the nucleus of an atom. It is made up of protons and neutrons. If we add or take away a neutron, it changes absolutely everything. It is no longer the same atom, and its properties will completely change. The lifespan of nuclear waste is fundamentally changed, and we could cut this from a million years to 30 minutes!*
>
> Gérard Mourou

There is a project that "may" operate within 10 years. Don't hold one's breath as there are lots of challenges. Not least is that operators of nuclear power plants have an entombment solution that absolves them of any long-term responsibility. Most companies will be long gone before they incur liability for their waste.

4.15.11 SMR Waste

Decommissioning SMR nuclear reactors will require the same duration and proportional cleanup costs. The NuScale deployment is intended to consist of 12 60-MW reactors, for a total of 720 MW of capacity. That suggests a cleanup cost range of $720 million. While I'm sure SMR advocates expect the reactors to be returned to a centralised processing site for decommissioning, no country in the world except Finland has managed to build a centralised nuclear waste repository, so this premise is highly unlikely and just a fantasy.

4.16 Accidents

Accidents happen. By their very nature, "Black Swan Events" cannot be predicted. Most reactor designs are such that if a system fails (e.g. cooling pump, rods), then the plant shuts down; otherwise, you would get a runaway reaction from the reactor. Pro-nuclear advocates say that deaths from nuclear power are the lowest of any electricity generation, but the key insight is that they are all much, much safer than fossil fuels. Safety has been a contentious issue since the 1950s. The risk of accidents has driven conservative practices into the industry.

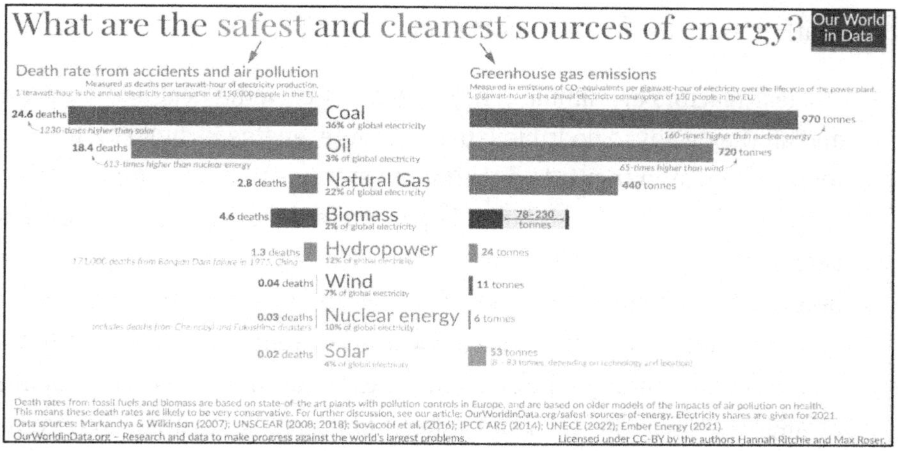

Figure 49 Safest and cleanest sources of energy

Figure 49 suggests that nuclear energy, for example, results in 99.9% fewer deaths than brown coal; 99.8% fewer than black coal; 99.7% fewer than oil; and 97.6% fewer than gas. Wind and solar are just as safe. These death rates are primarily from accidents and air pollution, and are very understated if the prediction of climate change impacts for fire, floods and famine come to fruition.

While we talked about "Black Swan Events" which cannot be predicted, most nuclear accidents are actually "Grey Swan Events". We can reasonably predict that an event is possible, even probable. We have a history of hundreds of accidents or near misses.

Nuclear can be viewed in a similar vein to an aeroplane. An aeroplane is heavy and takes a lot of energy to fly, and then all it wants to do is fall to the ground. It is pressurised so humans can fly without oxygen masks, but carries hundreds of tonnes of a lethal cargo of high-explosive petrochemicals right next to those passengers. It uses people as passengers and flight crew, and one in a million wants to do harm to other people. (Note – over 500 mass shootings in the USA with a population of 340 million) Despite all the engineering, redundant flight systems, backup flight system controls, and even more backup systems, along with pilot training, retraining, and recertification, they still fall out of the sky. We see doors fall off as the pins are missed. Or rogue flight copilots cut fuel supply at takeoff. All these accidents come with a tragic loss of life, but we trivialise those deaths due to the convenience of flying.

Nuclear reactors are similar. They operate at extreme heat and pressure, and all they want to do is blow up. Therefore, engineers build systems to monitor and stop them. And then more systems to make sure the safety systems don't fail. And build large containment vessels to prevent aeroplanes from crashing into them. We see nuclear power plants becoming weaponised in times of war, such as in Ukraine. Engineering systems built on top of other engineering systems introduce significant risks, primarily through the creation of cascading failures and single points of failure (SPOFs). These risks are inherent in what's known as a "system-of-systems" or "interdependent systems."

As of 2014, there have been more than 100 serious nuclear accidents and incidents from the use of nuclear power. Fifty-seven accidents or severe incidents have occurred since the Chornobyl disaster, and about 60% of all nuclear-related accidents/severe incidents have happened in the USA. [124]

4.16.1 What Is the Real Accident Rate?

Understanding the real accident rate is not easy. Rose and Sweeting [125] set out the arguments about nuclear safety, as there are two quite different approaches. One statistical approach is to examine the available data and calculate a probability. The other is to calculate the probability that nuclear is safe. Then use existing data to see how far away from zero safety is. The difference is, however, difficult to reconcile. One concludes that we could expect an accident every decade. The other says every million years. Which is the “more correct” approach? We will look at this in more detail as it is a key issue when accidents rates are mentioned.

Key Points About the Actual Rates:

- The authors analysed past core-melt accidents to estimate a failure rate of 1 per 3704 reactor years.
- That’s 1 per decade with the current 410 operating reactors.
- They found learning effects from past accidents were limited or absent, depending on the database used.
- The International Atomic Energy Agency (IAEA) does not have a publicly available list of nuclear accidents, making it challenging to analyse accident probability and learning effects.
- The authors used data compiled by the Guardian newspaper and energy researcher Benjamin Sovacool as an alternative source.

Data Insights:

- The study suggests that more severe nuclear accidents can be expected than previously thought.
- Nuclear power operators may not learn from past experiences, which could increase the risk of future accidents.
- The complexity and tight coupling of components in nuclear power plants make them inherently unsafe. Charles Perrow's "normal accident theory" posits that catastrophic failures are inevitable in certain types of high-risk technological systems. These aren't just random or rare events; they are "normal" in the sense that they are an expected and inherent property of the system's design.
- Statistical analysis supports the idea that serious accidents with complex high technologies are inevitable.

Implications from this Study:

- Public support for nuclear power cannot be based on full knowledge due to limited access to information.
- A more detailed analysis of nuclear accident probabilities requires transparency from the IAEA and better data availability.
- The study's findings highlight the need for caution when evaluating the safety of nuclear power plants, even with improvements in operation and reliability.

In reviewing major accidents such as Chornobyl and Fukushima we see information was suppressed, that authorities spun media to suit their interests and that most probably they were much more severe than the official lines.

4.16.2 Chornobyl Accident

The Chornobyl accident in 1986 [126] occurred during a safety test on the steam turbine of an RBMK-type nuclear reactor. A combination of operator negligence and critical design flaws had made the reactor primed to explode. Instead of shutting down, an uncontrolled nuclear chain reaction began, releasing enormous energy.

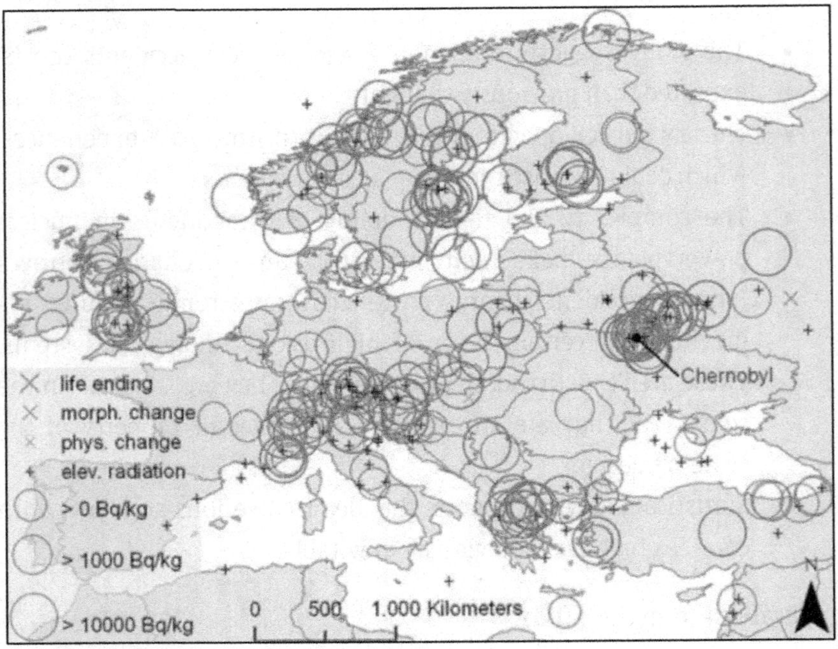

Figure 50 Spatial distribution of the impacts of the Chernobyl disaster described by the 521 studies included in the review

We have subsequently learnt that plant designers knew that this was possible – they never trained the operators. About 5% of the radioactive core was exploded into the atmosphere, and wind carried it across Europe. Figure 50 is shows the radiation 20 years later. The data is from 521 studies on ecological studies, which measured radiation levels. The size of the circle is the level of radiation at that particular site. Over 25% of radioactive particles are transported further than 2,000 kilometres. According to the official, internationally recognised death toll, just 31 people died as an immediate result of Chornobyl. At the time of the accident, the UN estimated that only 50 deaths could be directly attributed to the disaster. In 2005, it changed its predictions, stating that a further 4,000 might eventually die as a result of the radiation exposure. The Chornobyl disaster cannot be directly compared to atmospheric tests of nuclear weapons by simply saying that it is better or worse. This is partly because the isotopes released at Chornobyl tended to be longer-lived than those released by the detonation of atomic bombs. What do we know now about the accident forty years on?

- About 31 people died at the power plant. A further 600,000 liquidators (clean up people) were exposed. The government evacuated 115,000 people from the most heavily contaminated areas in 1986, and a further 220,000 people in subsequent years.
- Strong evidence links ^{131}Iodine exposure in children (<18 years old) to a linear increase in thyroid cancer risk, with a dose-response relationship of excess relative risk (ERR) per Gy of 1.91. [127], [128], [129]
- The International Agency for Cancer, WHO, say by 2065, models predict that about 16,000 cases of thyroid cancer and 25,000 cases of other cancers may be expected due to radiation from the accident, with over 16,000 deaths. [130]
- Risks persisted for decades without attenuation over time and were observed even at low doses. (e.g., ~0.05 Gy ^{131}I thyroid dose) [127], [128].
- Exposure to contaminated food and inadequate Iodine prophylaxis

were critical factors in amplifying risks. [128], [129], [131]

- Chornobyl accident caused US$235 billion in economic damages. [116]

We tend to focus on the people and have ignored other species. That review of over 521 papers following Chornobyl highlighted that elevated radiation levels were recorded among a diversity of species, even up to thousands of kilometres away from the meltdown site, and even after more than two decades. [28] Close to the reactor, physiological and morphological changes occurred with adverse effects on the ecosystem, including contamination of water, soils, and wild food supplies. Forty years on, the local environment appears to have partially recovered but little science work can be done currently due to Russia invading Ukraine.

4.16.3 Fukushima Accident

On March 11, 2011, the Tohoku tsunami flooded the Fukushima nuclear power plant.[132] The emergency pumps were at the bottom of the reactor and were not watertight, so seawater ingress stopped them. Once the pumps failed, the cores started to overheat, high levels of Hydrogen built up and then exploded. Poor training (operations) also contributed. The data about radiation levels and consequences depends on who is reporting – the authorities or independent researchers.

- Thyroid doses among the majority of Fukushima residents were low (most ≤10 mSv; maximum ~30-50 mSv) due to effective countermeasures, including evacuation and food controls. [133], [134], [135], [136]
- Large-scale systematic screenings through the Fukushima Health Management Survey (FHMS) detected high thyroid cancer prevalence, but the excess is attributed to overdiagnosis and screening effects, with no direct evidence of radiation-induced thyroid cancers [133], [137], [138]
- The attribution of "overdiagnosis" by TEPCO and UNSCEAR is at odds with best practice, which suggests that, with most cancers, early detection of cancers reduces fatalities. [139] [140] "Overdiagnosis" will have improved early detection and avoided

fatalities. There was a 30-fold increase in thyroid cancer cases in children and adolescents, with over 266 surgeries to remove their thyroid. These patients require lifelong ingestion of thyroid with a poorer quality of life, such as insomnia and fatigue.

- According to the 2020-2021 UNSCEAR report, the collective effective radiation dose to the population of Japan from the Fukushima reactor meltdowns over just the first ten years after the accident is 32,000 person-Sieverts. Multiplying this collective dose with the BEIR committee's estimates of cancer risk per unit of radiation exposure would result in roughly 1,800 fatal cancers. That Japan did not find and report this is allegedly a cover-up of data. You can't get data if you don't look for it.
- Clean up of land and plant is expected to take a further 30 more years and $76 billion to remove intact nuclear fuel, recover resolidified melted fuel debris, dismantle the reactors, and dispose of contaminated water. Removing the fuel debris is a tougher task, with no target completion date yet.

The close relationship between TEPCO (Tokyo Electric Power Company) and the Government directly caused the severity of the accident. The identified root cause was that the seawall was too low and emergency generators were not out of harm's way. Had it been built another 5m taller to deal with the 15m (1 in 1000) tsunami wave, water would not have flowed into the plant. The seawall was deemed too costly when they built the plant. At the planning phase, some engineers had recommended that the sea wall be taller to match the highest recorded in history. In 1986, the Government waived that engineering recommendation. Additionally, while the original site was 30m above sea level, they flattened the site to 10m because pumping water up to the 30m level would be more expensive! To compound the problem, the pumps were installed in the basement, not elevated, for aesthetic reasons. Worse, after Chornobyl, further reviews continued to ignore those previous tsunami records and took no action.

4.16.4 Comparing Chornobyl vs. Fukushima

The amount of radioactivity released at these two accident sites is known. Hindsight is easier. Note a PBq is a peta becquerel or 1 x 10^{15} or 1 followed by

15 zeros. It is a lot! When we look at the dosages people experienced around Chornobyl or Fukushima, we can see that these events are orders of magnitude different. Chornobyl released 10 times more than Fukushima. But Fukushima, due to better monitoring by the Japanese authorities compared to the Ukrainians, knows that 20% came from unit 1, 40% from unit 2 (peak on 15 March), and 40% from unit 3 (peak on 16 March). Releases to the ocean over 26 March to 30 September were about 11 PBq ^{131}iodine, 3.5 PBq ^{134}Cs, 3.6 PBq ^{137}Cs for a total radioactivity of over 18.1 PBq (or 169 PBq ^{131}I equivalent) apart from atmospheric fallout. The saving grace for Japan was the wind. More than 90% of the radiation went out to the ocean, not across the land. Had the wind moved the radiation across Japan with that amount of radiation, we would not be seeing Japan restarting its nuclear program or China developing its own, and there would have been thousands of additional cancers and deaths in Japan. These numbers can be confusing so here is the dose summary between Chornobyl and Fukushima:

- Chornobyl. Higher individual thyroid doses (often >100 mSv) (1,000 chest x-rays) due to delayed iodine prophylaxis and widespread contamination. [127], [128], [129]
- Fukushima. Lower thyroid doses (<30 mSv) (300 chest x-rays) primarily reflected early evacuation and containment measures. [133], [136], [137]
- Uncertainties. Fukushima dose reconstructions highlighted variability in early ingestion pathways for internal ^{131}iodine exposure, indicating areas for improved dosimetry models. [135], [136], [141]

The risks have not gone away. We have over 400 nuclear reactors of similar size. People still operate them, and people make mistakes. As the population increases, and as urban areas encroach closer to nuclear reactors, the consequences become more pronounced.

4.16.5 Three Mile Island

The accident on 28 March 1979 [142] occurred when the reactor was operating at 97% power. It involved a relatively minor malfunction in the secondary cooling circuit, which caused the temperature in the primary coolant to rise. This, in turn, caused the reactor to shut down automatically. Shutting down took about one second. At this point, a relief valve failed to close, but instrumentation did not detect it, and so much of the primary coolant drained away that the residual decay heat in the reactor core was not removed. The core suffered severe damage as a result. The operators were unable to diagnose or respond properly to the unplanned automatic shutdown of the reactor. Deficient control room instrumentation and inadequate emergency response training were the root causes of the accident. A 1997 study suggested people living near the Three Mile Island nuclear power plant, with the release of a small plume of radioactive gas into the air, may have suffered from a higher rate of some cancers in the first several years after the accident. Wing [143] reported people presumed to have been exposed to the highest doses of radiation were almost twice as likely to develop lung cancer as were those who received the lowest doses. His team also found that the risk of adult leukemia was nearly seven times higher for those in the highest exposure group. But other epidemiologists are sceptical of the claim of additional cancers. [139] The accident crystallised anti-nuclear safety concerns among the general public and led to new regulations for the nuclear industry. It accelerated the decline of efforts to build new reactors.

4.17 Nuclear for Australia

The anti-renewable lobby in Australia has taken up nuclear as an electricity supply, partly as anti-renewable and partly pro-fossil fuel. The conservative party, the Liberal National party, made building nuclear power plants a plank of its election promises in 2025. They took to the election a plan based on 8 nuclear power plants. That number was based on provision of the same electricity generation as the existing coal fired power stations generate.

Australia has no enrichment facilities, little skilled nuclear labour, not enough welders and has abundant wind and solar. Voters soundly rejected the LNP nuclear plant in March 2025, but the LNP Party appears determined to maintain a policy of building nuclear power, delaying renewable energy, staying with coal, and abandoning net zero.

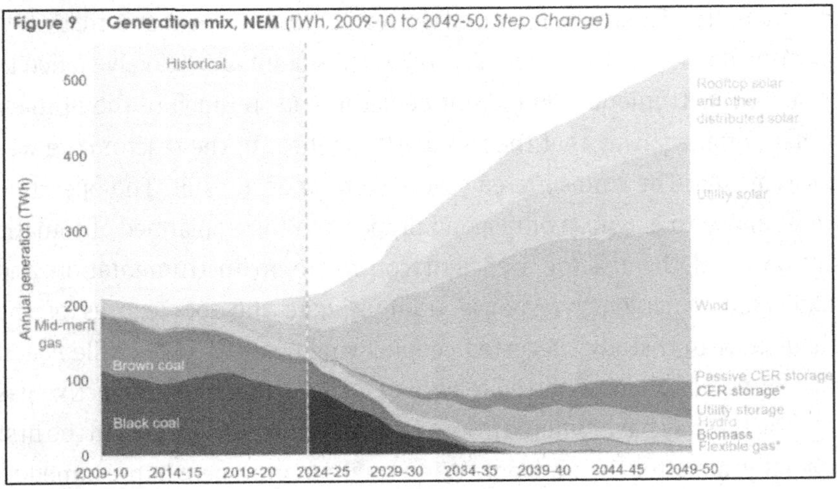

Figure 51 The generation mix under the AEMO Integrated System Plan shows double electricity consumption with large increases in wind and solar by 2050

Figure 51 is Australia's electricity grid plan. The Australian Electricity Market Operator (AEMO) publishes an Integrated System Plan (ISP) every 2 years, and has coal power stations retire by the end of 2035. By 2035, the grid would be 98% renewable energy and by 2050 electricity demand would be twice that of 2025. That is because electricity consumption doubles as it displaces household and industry gas use as well as increased demand with transport changing from petroleum and diesel to clean electricity.

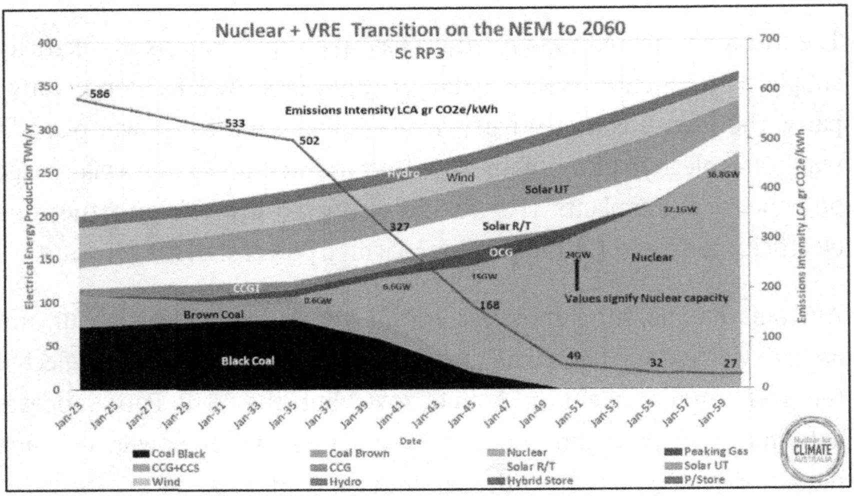

Figure 52 The nuclear pathway of Australia's nuclear plan increases emissions, delays the phase-out of coal until 2050.

Figure 52 produced by the pro-nuclear proposal was a stark contrast to Figure 51 produced by AEMO. In the renewable energy model of AEMO

- All new generation is wind and solar.
- Coal exits electricity generation by 2035.
- Total electricity doubles over 25 years to 500 TWh.

In the Pro-Nuclear Proposal by a political party that wants to stop renewable energy:

- No new wind and solar
- Coal plants continue to 2050.
- Nuclear, takes a decade until the first plant, then has to ramp up over 15 years.
- Total electricity demand only 30% increase to 270 TWh.

Politics aside, nuclear has no place in Australia. While Australia has uranium mines, Western countries' experience of 20 years of nuclear power plant construction that exceeds 15 years and costs over $10 billion per GW makes nuclear too slow and too expensive. Furthermore, Australia would need more than 30 nuclear power plants to produce 500 TWh of electricity. Politics ignore technology and economics.

4.18 Climate Change Risks

None of the 440 nuclear reactors built was designed with the impacts of climate change in mind. Even under the most optimistic scenarios (global net zero by 2050) a further 50cm sea rise is predicted by 2100. The Thwaites doomsday" glacier in the Antarctic could raise levels by 0.6m. With little action taken, we might see a rise of nearly 2.5 metres by 2100 and 5 metres by 2150. [145] In Part 1 of Optimistic Storm, in the Climate Change impacts section we know we will have rises in sea level. The only question is how much and when. As emissions rises, temperature rises and sea level rises are inevitable. How much? Uncertain.

Even a 1-metre rise is quite catastrophic. Five metres on that timescale would basically be an unimaginable disaster".

Dr Stefan Rahmstorf Uni Potsdam

Flooding can be devastating. It damages electrical systems, disabling cooling mechanisms, which can cause overheating, meltdowns, and the dangerous release of radioactivity. Over 25% of nuclear plants worldwide are situated in coastal areas; many were built only 10–20 metres above sea level when climate change was hardly seen as a threat. The Institute of Mechanical Engineers states that UK nuclear facilities may require significant investment to withstand rising sea levels or may eventually be abandoned or relocated.

4.19 Conclusion For Nuclear

Nuclear was going to be so cheap it would not be worth metering, but how wrong was that? What we are seeing is a progressive closure. Why?

- Electricity from nuclear is more than 3 times more expensive than other sources using existing plants.
- It takes too long to build new plants.
- Nuclear is a high-risk investment.
- Only built by governments.
- Provides low returns on investment.
- Is vulnerable to a lack of cooling water for generation.
- Is a terrorist risk (dirty bomb).
- Has low social acceptance.
- Has a limited Uranium supply
- Refinement in just a few autocratic countries.

The renewal rate with new plants is insufficient for nuclear energy to survive, says Schneider of the World Industry Status Report [65]. We have 411 operating reactors globally, but the forecast for 2040 is that no more will be built than will be retired. By 2030, let alone 2040, we will be in full-scale retreat. Some countries will continue their nuclear power plant fleets in conjunction with their nuclear weapon programs. Part of the focus on nuclear is that the IEA, other consultancy groups, the US Government and a few investors still believe nuclear has a role in 100% zero emissions scenarios. They have not changed for the past 20 years as wind, solar and batteries began their plunging costs. In many respects, it is like the fossil fuel companies that state there will be no decrease in the demand for fossil fuels by 2050.

Safety concerns aren't why nuclear is failing in the marketplace; economics are why nuclear is failing in the marketplace.
Michael Barnard *[146]*

The fuss with Prof Jacobson over his 100% renewable energy scenario that played out over the past 2 decades is still going on. Many reputable scientists bank their reputation on being pro-nuclear. Some went so far as to say that the safest place to be during the Great West Japan earthquake was on the boundary of the Fukushima Daiichi nuclear power plant, but in reality, that was the beginning of the end of the nuclear era. That was the fatal strike; solar and batteries are now delivering the finishing blows. Factors to consider include:

- Economics. Existing units should be closed as soon as possible to free up subsidies for investment in solar and batteries.
- Economics. Study after study, including Lazard [147] demonstrate it is cheaper to build new solar and batteries than to operate nuclear plants.
- Emissions. Some argue we need to continue to keep operating nuclear power plants to stop CO_2 emissions. When you assess the level of subsidies, that money would be better spent on expanding wind and solar ten times faster.
- Stop Funding Russia. We need to stop funding Russia for the enriched uranium.
- Waste. No one has sorted out a socially acceptable and cost-effective solution to the waste from nuclear power plants, weapons or medical waste.
- Waste. Keeping plants alive simply continues to increase the waste that will be with us for 100,000 to 1,000,000 years. Proponents who say waste is good and we can reuse it are seriously deluded. After 50 years, Australia still can't even agree on where low-level waste can be stored. Entombment is the only method today.

Here is what we should do with nuclear.

- Proposed new builds. Stop planning now.
- Sites under construction. Stop. Abandon.
- Put the future expenditure towards domestic solar and batteries to decarbonise as fast as possible.
- Continue some research on fusion and thorium.

Amid the European gas crisis, France backed off plans to close 14 nuclear plants, but plans to build more have soured due to economic reasons. For existing nuclear plants in operation, assess whether the subsidies are worthwhile; if so, refurbish and maintain. That will only be for relatively modern plants. For those who require large subsidies to keep operating, close. The government's focus needs to be single-mindedly on reducing CO_2 emissions. Politics is very clear. Investing in future nuclear power is used to avoid investing in wind, solar, and batteries. While greed rules, stupid people will grasp at nuclear. Any new nuclear plant started today will not come onstream until late 2030, and by then, it will be hopelessly uncompetitive. Electrons from solar wind and batteries spell the end of fossil fuels. And also spell the end to nuclear. The faster the better.

Conclusion of Nuclear:

- Nuclear costs 3 times more and is a sunset industry compared to wind solar and batteries.
- New plants take decades to build.
- Waste is not dealt with. Pretending otherwise is criminal to future generations.
- Accidents are ever-present.
- Military objectives are driving civilian investment.
- How false narratives are driving baseload.
- Nuclear fusion generators, SMR small reactors, fast breeder reactors, and thorium reactors are attempts to patch over fundamental issue. It is molecular energy, not harvesting energy.
- The number for Australia to replace existing coal generators and supply electricity to 2050 is over 50 units.
- We would need 2,000 SMR reactors just to keep up with new wind and solar – in 2024 alone. 3,000 in 2025.

5 Carbon Capture and Storage

Explore Carbon Capture and Storage:

- What is CCS and its status?
- Purpose of CCS?
- Failure to scale?
- Why do companies and governments persist?
- What will be the likely status in 2 decades?

Carbon capture and storage process (CCS - sometimes CCUS with a U for underground) aims to mitigate climate change by capturing carbon dioxide emissions from industrial processes, power generation, or other sources, and then storing them in underground geological formations or other secure storage locations. The purported purpose of CCS is to prevent the release of CO_2 into the atmosphere, a major contributor to global warming and climate change. The reality is CCS is a scam concocted by the fossil fuel industry to delay stopping the burning of rocks (fossil fuels).

"When an energy technology is consistently shown not to work, but can attract subsidies, you can be sure there are lobbyists and greenwashing at play. The oil and gas industry are harvesting subsidies to prolong the life of their operations, not for emissions reduction.

Bruce Robertson. IEEFA

5.1 CCS Status

Despite the technology being available for over 50 years, CCS is almost non-existent.

- No natural gas-fired electrical power plant in the entire world currently uses CCS. None are expected to open until after 2025 [148] and natural gas combustion is expected to require greater energy input than coal combustion to achieve a high capture rate. [149]
- Only one coal-fired power plant in the world currently operates with CCS. The Petra Nova project in Texas, suspended operations in 2020.
- Only two commercial plants producing hydrogen from natural gas capture more than 1 million metric tons per annum (Mtpa) of CO_2–one in the United States (Air Products' Port Arthur, Texas, facility) and one in Canada (Quest).
- No new CCS-equipped plant producing hydrogen from natural gas and capturing 1 Mtpa or more of CO_2 is expected to open until 2030 or later.

5.2 How CCS Works

The CCS process typically involves three steps: capture, transport, and storage.

There are several capture technologies, including pre-combustion, post-combustion, and oxyfuel combustion. The 50-year history of mechanical carbon capture efforts is entirely tied to the fossil fuel industry, most of which are used to pump more oil out of played-out oil wells as enhanced oil recovery approaches.

The CO_2 has to be transported to the injection site – usually by pipelines and then injected into the rock formation.

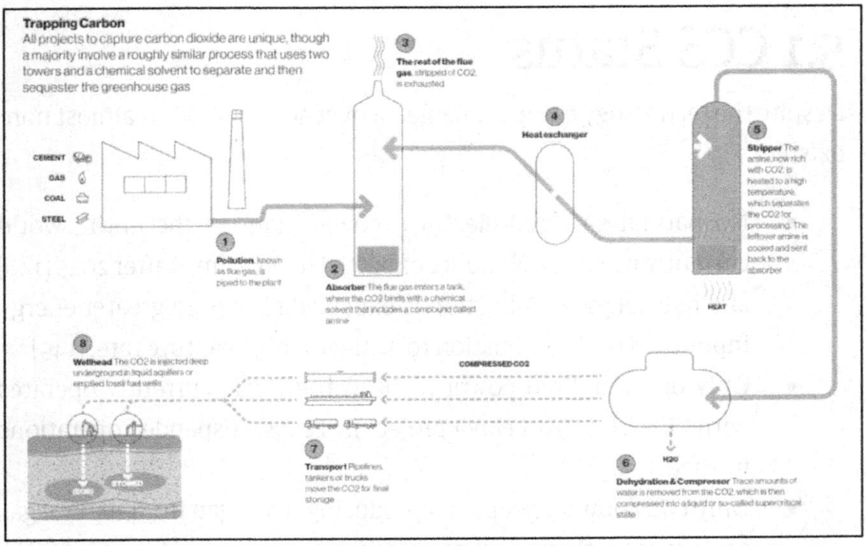

Figure 53 Carbon capture and Storage is complex, expensive, and a scam.

Figure 53 shows the complex process of capturing CO_2 from a waste stream in a coal, cement, gas or steel plant. CO_2 is captured from industrial processes or power plants before it is released into the atmosphere. Direct Air Capture tries to retrieve CO2 from the atmosphere.

5.3 CCS Is Ineffective

CCS reduces only a small fraction of carbon emissions, and it usually increases air pollution.

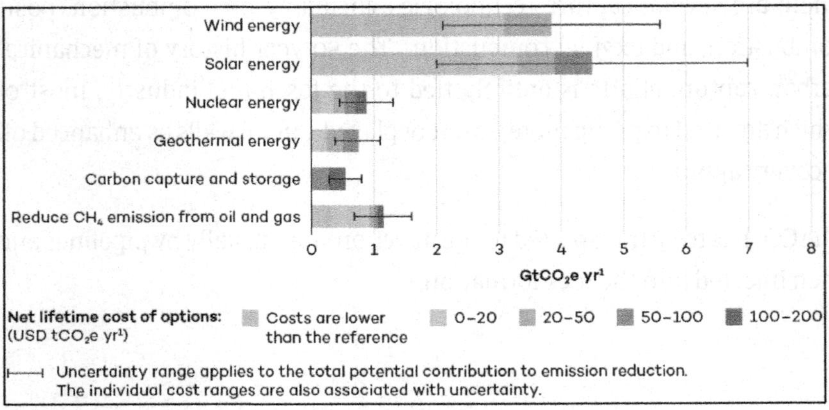

Figure 54 Carbon capture and storage vs other technologies for lifetime reduction of CO_2

Figure 54 shows that reducing CO_2 with wind and solar is the most effective and can reduce CO_2 by 2 to 7 Gt per year. In contrast, CCS is unable to contribute more than about 0.5Gt, and progress to date is a fraction of that. Even if you have 100% capture from the capture equipment, it is still worse, from a social cost perspective, than replacing a coal or gas plant with a wind farm. CCS never reduces air pollution and always incurs capture equipment costs. Studies in Canada, Australia and Norway concluded CCS is worthless and that directly reducing emissions using wind or solar is orders of magnitude less costly and more effective.

Nuclear energy is also a poor option to reduce emissions. That is not due to the reduction contribution once the plant is operating. The long delay to build and the cost of construction are huge opportunity costs. Nuclear energy is estimated to be five times less cost-effective than wind and solar.

Jacobsons assessment showed that, by global standards, virtually no CO_2 has been captured by CCS, and that a single year's output from current wind and solar farms avoids 35 times the CO_2 that has been 'captured' over 50 years of CCS history. All CCS is a rounding error on the actual solution, and we need to just stop emitting CO_2. [150]

5.3.1 Norway CCS Case Study

The Norwegian Equinor (nee Statoil) Sleipner facility in the North Sea stands out for emitting 25 times more CO_2 than it captures. Michael Barnard calculates [151] that when CCS is used to extract more gas, which, when burnt, the net effect is CCS produces 25 times more CO_2 than leaving the gas in the ground. The analysis showed:

- CCS purpose was that the LNG had 9% CO_2, too high for sale of the LNG.
- Norway had a carbon tax, so Equinor saves on carbon emissions for scope 1 and 2.
- Removes CO_2 and sequests it underground.
- About 1 million tonnes stored per year – 23 million tonnes since 1996 to 2019.
- For every 1 tonne of natural gas, take 90kg CO_2 and sequester it.
- The balance of the tonne of gas, 910 kg, is burnt by customers,

produces 2,500 kg of CO_2.

- Therefore, net CO_2 is 2410 kg for every 1 tonne of gas extracted.
- Produces 36m^3 per day, and has been producing that since 1996, 300 billion m^3, or 581 million tonnes CO_2, compared with 23 million tonnes CO_2 sequestered.

In 2025, Equinor [4] retracted its claim that it stores about a million tonnes of carbon dioxide annually at its flagship carbon capture project after DeSmog obtained data showing the real figure was as little as a tenth of that amount.

> *That's over 25 times more CO_2 in the atmosphere than was sequestered. And Equinor is being paid for the natural gas and the sequestered CO_2. Nice work if you can get it. Not so nice for the planet.*
>
> Michael Barnard, CleanTechnica *[150]*

5.3.2 Canada CCS Case Study

An International Institute of Sustainable Development report [152] on Why Carbon Capture and Storage Is Not a Net-Zero Solution for Canada's Oil and Gas Sector highlights the challenges that Canada faces. Reducing emissions from Canada's oil and gas production is a priority, yet fossil fuel industry representatives consider CCS the sector's primary emissions-reduction solution.

There are seven CCS projects currently operating in Canada, mostly in the oil and gas sector, capturing about 0.5% of national emissions. CCS in oil and gas production does not address emissions from downstream uses of those fuels. The captured carbon is used predominantly for enhanced oil recovery to facilitate additional oil extraction.

CCS in the oil and gas sector is expensive—as much as $CAD200 per tonne for currently operating projects—as well as energy intensive, slow to implement, and unproven at scale, making it a poor strategy for decarbonising oil and gas production. Despite this, the federal government provides substantial support for CCS, having committed at least CAD $6 billion to date, alongside $2.6 billion from the governments of Alberta and

Saskatchewan. Industry is seeking further public funding. By 2030, Canada's CCS Investment Tax Credit is estimated to provide double the subsidy amount offered via the IRA in the USA, although the USA CCS Subsidies in the IRA were rescinded. Do taxpayers get a return on the subsidies paid. No.

5.3.3 Gorgon CCS Project in Australia

The Gorgon Carbon Capture and Storage (CCS) project is located off the coast of Western Australia. Chevron, in partnership with ExxonMobil and Shell operate the CCS plant. It was part of construction approvals.

- The Gorgon gas development extracts methane gas from offshore fields.
- Processes gas on a nearby island.
- The gas is piped to onshore facilities for liquefaction. About 15% of the gas is used to liquify the gas and then the gas is exported.
- CCS plans to capture and store up to 4 million tonnes of carbon dioxide (CO_2) per year.
- CO_2 equivalent to the emissions from around one million cars.

5.3.3.1 Gorgon CCS Reality

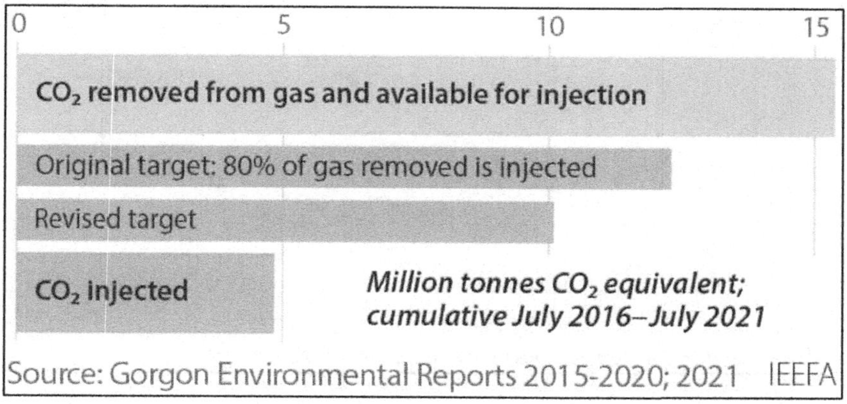

Figure 55 Australia Gorgon CCS fails to meet its target by 50%

The reality as shown in Figure 55 is that of the gas removed, less than a third can be injected.

5.3.3.2 CO_2 Storage

- CO_2 is captured during the processing stage, prior to the liquefaction plant.
- Captured CO_2 is transported through a 225 km-long pipeline to the injection site.
- The injection site is located approximately 3 km beneath Barrow Island, a Class A nature reserve. The injection site is two separate storage reservoirs, the Dupuy Formation and the Triassic Formation.
- Initial assessment of the deep saline aquifers suggested suitability for long-term CO_2 storage.
- The injection process involves pressurising the CO_2, pumping it into the reservoirs through a network of wells.
- CO_2 is stored in the reservoirs in a supercritical state.
- The CO_2 is in a dense, liquid-like form held in place by the weight of the overlying rock formations.

Figure 1: Gorgon CCS Plant's Performance (Cumulative Trend, 2016-2021)

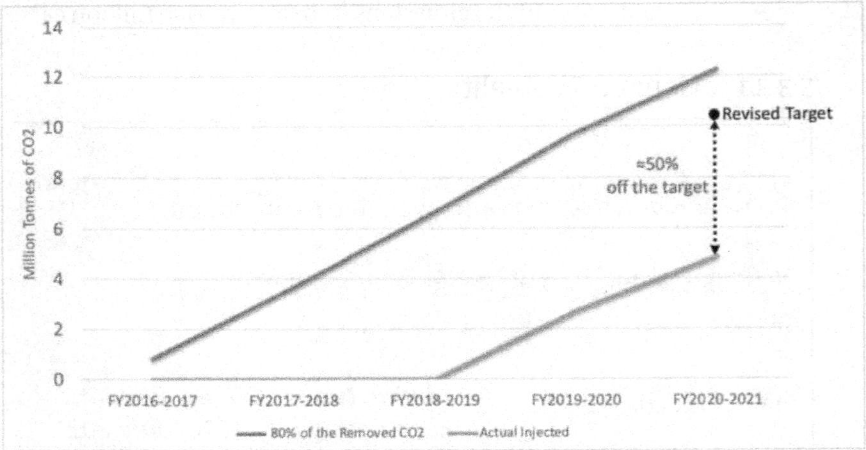

Source: Gorgon Environmental Performance Reports 2015-2020 & 2021.

Figure 56 Gorgons CCS Plants Performance cumulative trend of forecast and actual to 2021

5.3.3.3 Reality of Project

- Cost of project = estimated $ 3 billion
- The carbon is only Scope 1 and 2 emissions, not Scope 3. Scope 3 is from gas usage.

- Only designed to send CO_2 to the reservoir for CO_2 removed during processing at the gas treatment plant.
- Design to sequester 80% of CO_2 (4 million tonnes pa) over a 5 year period from 2016
- Failed to start until 2019.
- Target performance was to reduce greenhouse gas emissions by 40% (100mt).
- Operation has engineering issues, including sand, water, or the carbonic acid it forms with CO_2. As well as corroding transport pipelines it damages the equipment in the injection well.
- Gorgon claims it is currently operating at around 80% of its planned capacity. However, 2021, it operated at just half its capacity at 2.2 million tones compared to the planned 4 million tonne capacity.
- Has stored less than 1% of the emissions from the total Scope 1, 2 and 3 emissions.
- Australian taxpayers are liable for the project post-closure.

When one stands back and looks at the bigger picture, the Gorgon Gas Field is one of the world's largest emitting industrial facilities, generating approximately 9.4 million tonnes of Scope 1 greenhouse gas emissions annually. It also produces more than 50 million tonnes in Scope 3 emissions when the gas is ultimately burned by customers. CCS captures less than 2% of emissions. The revenues from the Gorgon project are estimated to be over $440 billion over its lifetime. With no penalties for increasing climate change. That money spent on CCS would have far better returns building wind and solar to remove coal and gas from the electricity grid.

5.3.4 USA Petra Nova Study

Petra Nova is a large-scale carbon capture and storage (CCS) project located at the W.A. Parish Generating Station in Thompsons, Texas and designed to reduce CO_2 emissions from one of the coal-fired boilers at the coal plant. It captured about 33% of the emissions from unit 8. The captured CO_2 is piped 100 km to the West Ranch oil field where it is injected to boost oil production from 300 to 15,000 barrels of oil.

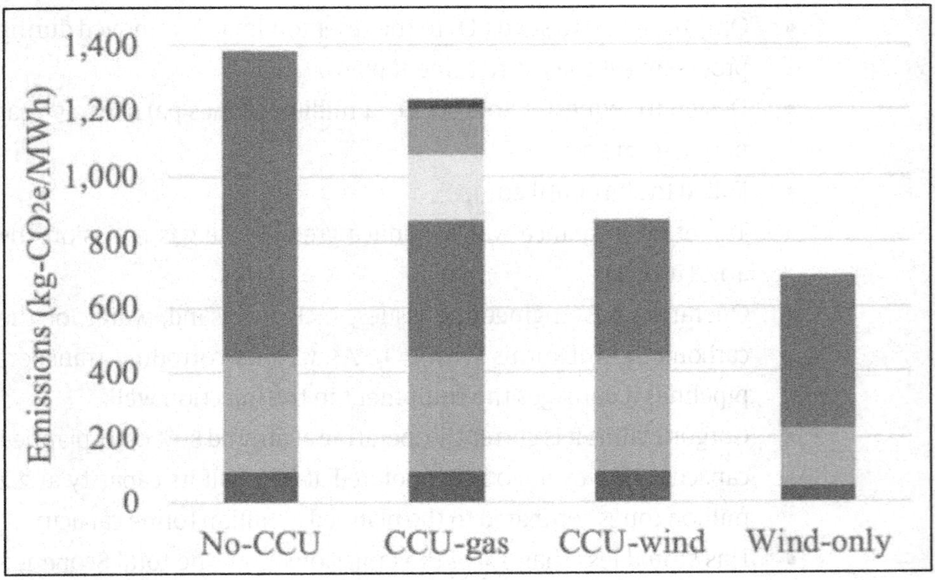

Figure 57 CO_2 emissions, averaged over 20 years, from the Petra-Nova coal plant from Jacobson

In Figure 57, Jacobson [153] calculated the emissions from each part of the process. He compared four different strategies. (a) Unabated burning coal. (b) Burning coal. Using gas to power the CCS CO_2 sequestration process. (c) Burning gas but using wind to power the CCS process and (d) using wind for electricity and using some unabated coal to make up the same electricity generation. The study results mirror those of other CCS projects that CCS is worse than doing nothing at all.

- The no-CCU column shows emissions from coal. The bottom blue is the coal mining emissions – CO_2 and not methane (CH_4). The orange are emissions from CH_4, and the major red portion are emissions burning the coal.
- The CCU-gas column shows emissions after the addition of the CCS equipment. It is powered by natural gas. The emissions are very similar to just burning coal.
- CCU-wind is similar to the CCU-gas, but instead of using gas to power the plant, wind is. Emissions are down 30%.

The lowest emissions were achieved by using wind and burning some coal without CCS to meet the total electricity output.In conclusion, the CCS only added a billion in costs, was ineffective and the emissions were 50% of coal.

In 2022, NRG Energy sold its 50% stake in this plant for about $3.6 million, far less than the $1 billion construction costs. The US DOE had sunk about $195 million in the plant, hoping to demonstrate its potential. The actual CO_2 capture rate was never published, apart from Jacobson's study and was substantially lower than 90%, perhaps as low as 65% to 70%. This firesale demonstrates the technology and economic failure of CCS. Just deploying wind achieved twice the reduction in emissions than CCS.

5.4 Forecasts By IEA And Others

Forecasts by various consultancies [154] are so far off the mark as to be ludicrous. There are 35 operational/operating CCS plants globally, most using CO_2 to enhance end-of-life oil fields.

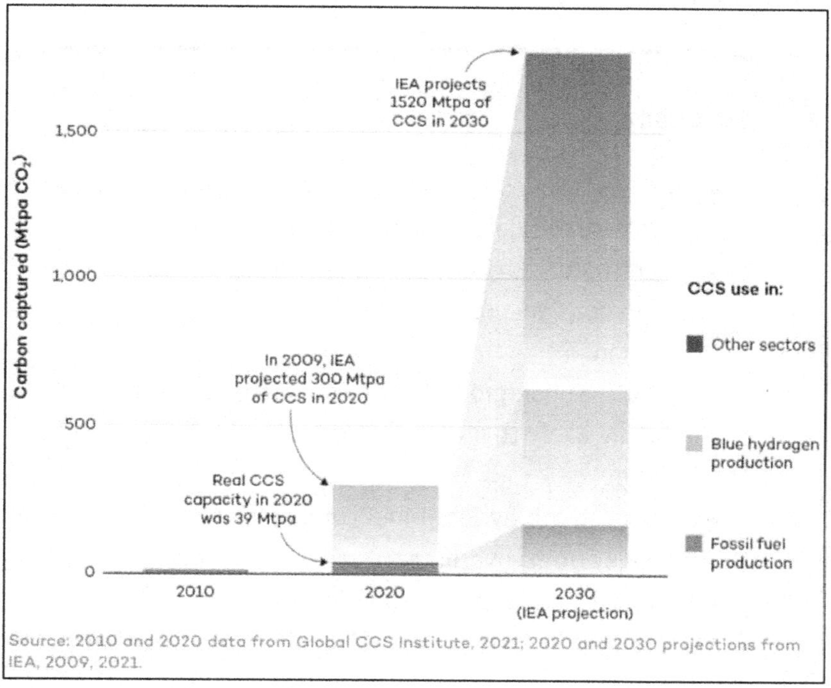

Figure 58 Forecasts from the Global CCS Institute 2021 with projections from IEA 2009 and 2021

We have looked at four case studies and show that the fossil fuel companies have managed to extract large grants and subsidies from governments, make promises, but deliver less than 10% of forecasts. In Figure 58, one has to assess how the IEA arrived at 1,500 million tonnes of CCS by 2030 at a cost of over $4.5 trillion when the current amount was less than 39 million tonnes, and some of those have already shut down. In the last three decades, governments and corporations have poured $83 billion into projects. In 2022, the technology captured less than 0.1% of global emissions.

5.5 Conclusion: CCS

CCS is a technology used by oil companies to extract more oil and to capture politicians and suck out vast amounts of taxpayers' dollars. . While virtually no CO_2 by global standards has been captured by CCS, a single year's output of wind and solar farms in 2022 avoided 35 times more than the past 50 years of effort. By 2025, we see renewable energy meet all the increased electricity demand and increasing by over 35% year on year.

Conclusion of CCS:

- CCS is a failed technology. The majority of use cases is t increase the yield of oil extraction.
- CO_2 captured by CCS in the past 30 years is less than 0.1% (emissions yet companies and governments invested over $8 trillion.
- Storage underground and associated structures suggest it highly likely that escape CO2 will bring its own issues in th future.
- It is con job by fossil fuel companies to continue to extra money from government.
- Stop CCS and capturing politicians.
- Building more wind and solar is 70 times more effective tha CCS.

6 Chapter 5: Direct Air Capture

Explore Direct Air Capture:

- What is DACS and status?
- Why does the technology fail?
- Why do companies and governments persist?
- What will be the likely status in 2 decades?

Is direct air capture (DAC) a promising technology or a scam? DAC is a technology that removes carbon dioxide (CO_2) from the atmosphere. DAC systems are considered by many to be simply greenwashing. It is still in the early stages of development, and the cost of CO_2 capture remains high. Costs should decrease as technology improves, making it a viable option for reducing greenhouse gas emissions. The key cost driver is energy. To be low emission, that energy needs to be renewable. Here are some of the benefits of direct air capture:

- Reduce the concentration of CO_2 in the atmosphere. Plants do.
- Capture CO_2 from a variety of sources, including power plants, factories, and vehicles.
- Produce high-purity CO_2 for use in other applications, such as food and beverage production, and the production of synthetic fuels and chemicals.

Conversely DAC has challenges including the following:

- The technology is still in its early stages of development.

- The cost of capturing CO_2 is very high.
- Shell predicts the energy needed is 100EJ for carbon-sucking machines to absorb more than 5 billion tons of CO_2 per year. [155]
- The energy required to operate DAC systems can be a significant source of emissions in itself.
- Environmental impact is poorly understood.

Overall, DAC is a promising technology for reducing greenhouse gas emissions. Before wide deployment, there must be major reductions in costs and energy use. Any DAC program should be only research. It is not ready for widescale deployment. Even more important, reducing CO_2 by stopping the burning of fossil fuels today is the most important. When we have switched out all energy sources and stopped pouring more CO2 into the atmosphere, it will be time to look at how to reduce CO_2 concentrations. As the saying gones: when you are in a hole, stopping digging.

6.1 DAC Methods Systems

DAC systems use a variety of methods to capture CO_2, including:

- Chemical absorption: A liquid solvent is used to absorb CO_2 from the air. The solvent is then heated to release the CO_2.
- Physical adsorption: A solid material is used to adsorb CO_2 from the air. The material is then heated to release the CO_2.
- Electrochemical separation: An electrochemical process is used to separate CO_2 from the air.

6.2 USA Funding

The U.S. Department of Energy (DOE) announced [156] it would spend $1.2 billion on two DAC facilities—one in Texas and the other in Louisiana – to remove millions of tons of carbon dioxide (CO_2) annually from the atmosphere. Part of a controversial effort to combat global warming, the awards represent the first phase of $3.5 billion in funding for DAC hubs set aside in the 2023 Bipartisan Infrastructure Law and marked the first major governmental backing in the world for the emerging carbon capture technology.

The program also aimed to create four DAC hubs over the next 10 years, each capable of removing and storing at least 1 million tons of CO_2 each year. As part of the program's rollout, DOE officials also announced funding for an additional 19 conceptual and engineering studies of potential future DAC plants. Those smaller projects were cancelled by the Trump Administration but the 2 major ones retained funding as the Secretary of Energy stated they "had a credible way to be □helpful." [157] Given the technology and economics, how they will be helpful is not entirely certain.

6.3 DAC In Iceland

Direct air capture in the first notable DAC plant in Iceland, built by ClimeWorks, has failed to cover its own emissions. [158] It has captured only just over 2,400 carbon units since it began operations in the country in 2021, out of the 12,000 units that company officials have repeatedly claimed the company's machines can capture. In 2024 Climeworks partially commissioned the Mammoth capture plant, which is expected to capture nine times more than what had been done since 2021, or 36,000 tons. That plant has managed to capture only 105 tons of CO2 in its first 10 months of operation. That does not even cover the carbon cost of building the plant.

The Swiss founders of Climeworks were ambitious when they started in 2009. In a 2017 interview, they said that by 2025, they planned to capture one per cent of all global emissions. That amounts to 400 million tons of CO_2. Those plans have not been realised, and the company has never come close to achieving that. They also planned to reduce the cost of capturing each ton of CO_2 from the atmosphere to about $100. Today, a ton of CO_2 costs about $1,000, according to the Climeworks website – ten times more than the target this year.

6.4 DAC Research

Prometheus [159] is one company looking to cut costs and has announced a low-cost methanol using direct air capture [DAC] and intermittent renewable energy, fully off-grid. The process uses a solution of liquid water and CO_2, which is exposed to an electrically charged copper plate. This catalyses a reaction and produces ethanol. Closely packed filters made from cylindrical carbon nanotubes embedded in plastic allow ethanol through while blocking water molecules. The more concentrated solution of approximately 95% ethanol can be catalysed with zeolite to form e-fuels such as methanol. They call it the Prometheus Faraday Reactor. At 50 cells, the Faraday Reactor is the full-scale unit in the process. They claim pricing below $50/tonne, which is 2.5 times the technology's theoretical bottom limit. If this were true, it would change everything we know about the DAC and subsequent decarbonization pathways overnight. They are scheduled for completion in 2025. [160] Their process bypasses traditional gas purification, compression, absorption and desorption, and costly infrastructure -dramatically reducing both energy use and capital requirements. By early 2026 they announced a "Prometheus eKerosene pathway that would make the Fischer-Tropsch synthesis obsolete. It would reduce the cost to make e-Fuels by more than 80 per cent compared to the century-old technology. It is also the first process in history to make kerosene from electricity and atmospheric CO2 without using hydrogen. The kerosene produced is 100 percent carbon neutral." [1]

6.5 Conclusion: DAC

DAC may have a role in reducing CO_2 in the air. It may enable f-fuel production at a price that H_2 cannot. However, it is like trying to suck up CO_2 with a straw when a major river of CO_2 is being emitted by the 100 companies that account for 71% of emissions.

Instead of trying to expand DAC, we need to do the following.

- Remove subsidies of over $13 trillion and 7% of global GDP currently paid to emitters.
- Reallocate to the 100 companies that have clean technologies to accelerate the deployment of clean energy technologies. That will meet the increasing electricity demand and push out CO_2 and methane energy sources.
- Stop giving carbon emitters a voice. Just as we did with tobacco companies.
- Turn attention to agriculture, which accounts for 20% of emissions, and increasing areas of forests are being destroyed to make way for

ever-increasing production of biofuel and other non-food products. Forests and grasslands remove CO2 at no cost and with no large renewable energy inputs.

- In a decade, come back to DAC and assess its economic value.

Conclusion of Direct Air Capture

- The Challenge – how to reduce CO_2 concentration in the air.
- Solution is rapid decarbonisation to stop more CO_2 emissions.
- Replanting 70% of the land area currently used for agriculture could remove most of the air CO_2 within a couple of decades.
- Removal of the carbon in the ocean is a problem that will haunt civilisation for centuries

7 Biofuels and E-Fuels

Explore Biofuels and E-Fuels:

- What are biofuels (ethanol) and e-fuels (methanol)?
- Sources of these fuels.
- Use as low emissions fuel in transport instead of oil using existing internal combustion engines.
- Why are companies and governments persisting to push biofuel and e-fuels?
- What will be the likely status in 2 decades?

Understanding ethanol, methanol, biofuels, and e-fuels can be confusing, yet the terms get thrown around as if everyone were a chemist. Biofuels and e-fuels are often touted as renewable alternatives to fossil fuels. Still, their fundamental difference is how they're created: one from living matter and the other from a chemical synthesis process powered by emission-free electricity. Methanol and ethanol are both alcohols but have distinct chemical structures, properties, uses, and effects. Ethanol is found in alcoholic beverages, but methanol is highly toxic and must never be consumed. Both have functional hydroxyl (-OH). Ethanol is a two-carbon molecule, C_2H_5OH. Methanol is a single-carbon molecule, CH_3OH.

What a comparison of synthetic ethanol and methanol highlights is that both have the potential to reduce carbon emissions, compared to their traditional fossil fuel counterparts. However, their production processes, feedstocks, and overall sustainability profiles differ significantly. Today, over 90% of ethanol is produced by fermentation, while methanol is produced by a synthetic process that steam reforms natural gas to create synthesis gas. Moreover, methanol has a higher energy content than ethanol, and, combined with industrial factory production using methane gas, this means a larger market for fossil fuel-based methanol. As we will explore, low-emissions methanol (e-fuel) needs a different manufacturing process. Ethanol is generally not used as a primary energy source, but blended with gasoline (E10 mixes) in internal combustion engines.

7.1 At A Glance

While the end product may be the same, with a fuel such as Sustainable Aviation Fuel (SAF), the source and process may differ.

Feature	Biofuels	E-Fuels
Primary Ingredient	Biomass (Crops, waste, algae)	Electricity, Water, and Captured CO_2
Land & Water Use	High. Requires land/water to grow crops.	Extremely low
Production Cost	Relatively low (established technology)	Very high (expensive technology)
Scalability Limit	Limited by available arable land/waste	Limited by the availability of clean electricity

Figure 59 Biofuel and E-Fuels differences at a glance.

In Figure 59, we can see that biofuel is a long-established fuel source, whereas e-fuels are recent and rely on emissions-free electricity. More detail follows.

7.2 E-Fuel

E-fuels (electrofuels) are synthetically produced. This is a chemical process, not a biological one. They need low-carbon electricity. They're considered "drop-in" fuels because they are chemically similar to fossil fuels and can be used in existing engines and infrastructure without significant modifications.

7.2.1 The Syngas Process

Feedstock. Carbonaceous feedstocks (like coal, natural gas, biomass, or municipal waste) are gasified to create CO and H_2

Catalytic Synthesis: The syngas is passed over a transition metal catalyst (typically iron or cobalt) at high temperatures , 150-300 °C at pressure.

Polymerization: The CO and H_2 undergo a polymerization-like reaction on the catalyst's surface, called the Fischer-Tropsch reaction linking together to form long-chain hydrocarbons and water.

$(2n + 1) H_2 + n CO \rightarrow C_nH_{2n+2} + n H_2O$.

Most of the hydrocarbons are alkanes – useful for diesel and similar applications. Methane produced in this process is not very useful, aside from lower production emissions compared to natural gas methane.

7.2.2 Primary Ingredient - Feedstock

Synthetic Methanol is primarily produced from syngas (synthesis gas), a mixture of hydrogen and carbon monoxide. It is a catalytic process in which various catalysts and reaction conditions influence yield and energy efficiency. After the formation process, distillation and purification separates methanol from other products and impurities.

Syngas can be derived from:

- CO and H_2
- Biomass: Sustainable option using wood, agricultural waste, or algae. It is more common with ethanol than with methanol.
- Coal or Natural Gas: Unsustainable due to high carbon emissions.
- Synthetic Ethanol using bacteria
- Direct Air Capture (DAC): Cutting-edge method capturing CO_2 directly from the atmosphere. See previous chapter on DAC.

- R&D may find other ways to create methanol from coal without emissions[162] but it is still in the very early research days.

Fossil fuel sources - the traditional method involves the steam reforming of natural gas or coal, resulting in significant CO_2 emissions.

7.2.3 E-Fuel Types

E-fuel types include:

Name	Chemistry	Use
e-Gasoline	Synthetic isooctane (C_8H_{18})	A "drop-in" replacement for traditional gasoline in vehicles with internal combustion engines.
e-Diesel	Synthetic long-chain hydrocarbons	A "drop-in" replacement for fossil diesel in vehicles, heavy-duty trucks, maritime shipping, and rail.
e-Kerosene	A lighter form of diesel. Jet-A or Jet-A1	Jet fuel is essentially a highly refined, lighter kerosene. Diesel is heavier and formulated with specific additives.
e-Methane	Methane (CH_4)	Can be injected into existing gas grids. Heating, industrial processes, and as a fuel for heavy-duty vehicles and ships.
e-Methanol	Methanol (CH_3OH)	Shipping, power generation, and a feedstock for the chemical industry. It can also be a stepping stone for creating other e-fuels.
e-Ammonia	Ammonia (NH_3)	Maritime shipping, power generation, an industrial chemical (e.g., for fertiliser production).

Figure 60 E-Fuels Type, usage

Figure 60 shows the end products. Existing methanol dominates the market but is unsustainable for long-term climate goals. It is cheap at about $0.30/kg. The "green" methanol processes above rely on hydrogen production and therefore cannot meet that price. Capturing carbon emissions using CCS won't be cost-effective – see Chapter 5 on CCS failure.

Today, production costs for e-fuels are high due to first-of-a-kind e-fuels small production plants. The biggest contribution to cost is prohibitively expensive green H_2. Over time, the costs of producing e-fuels are expected to fall significantly as experience, technological progress, and scaling increase, similar to prior experience with renewable energy and electricity storage. Frontier Economics estimated long-term production costs (including transport to Germany) for e-petrol at €0.99-1.63 per litre and for e-diesel at €1.09-1.80 per litre in 2025. They got it wrong.

Many get enthusiastic because they do offer a solution to hard-to-abate emission industries, such as long-haul aviation, which makes up about 3.1% of global emissions.

7.2.4 Biomass-Based (Bio-methanol)

Bio-methanol requires sustainable biomass sources like wood chips, agricultural waste, or algae to produce syngas (synthesis gas), which is then converted to methanol. While emissions are lower than those from fossil-based methods, some emissions still occur during biomass growth and processing.

7.2.5 DAC for e-Methanol

This cutting-edge method captures carbon dioxide directly from the atmosphere and combines it with hydrogen produced from renewable electricity using electrolysis. The resulting syngas is converted to e-methanol. The process offers potential for negative carbon emissions if combined with efficient carbon capture and storage. See the section in DAC in Chapter 6.

7.3 E-Fuels for Land Transport

The efficiency of biofuel and e-fuels for transport are bad. The IEA has an assessment of e-fuel to decarbonise transport [163] and is remarkably bullish, stating rapid deployment of low-emission fuels during this decade will be crucial to accelerate the decarbonisation of the transport sector. IEA says that 10% share of e-fuels in aviation and shipping by 2030 is feasible, but it would need a massive build-out of renewable energy.

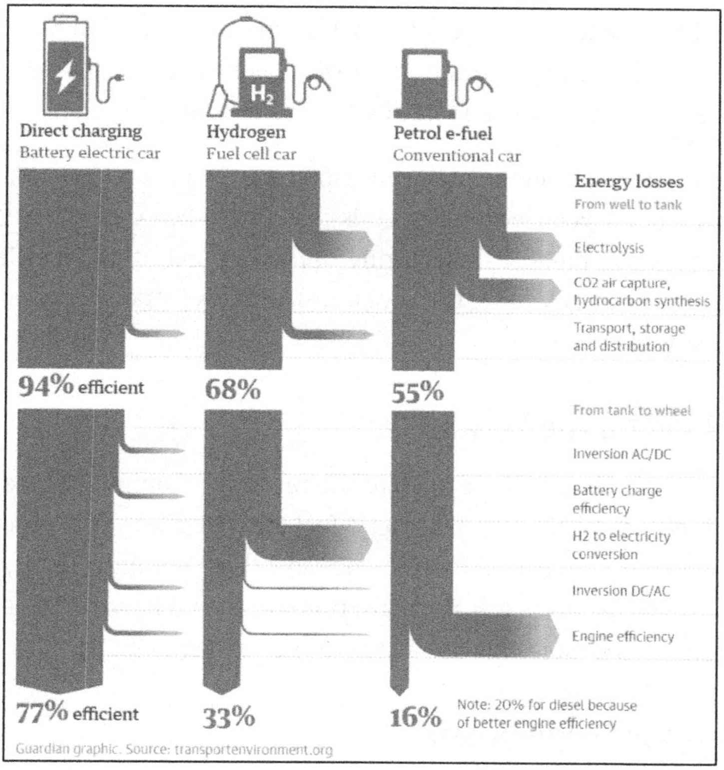

Figure 61 Comparison of losses in energy for petrol, hydrogen and e-fuel in motor vehicle by Transport and Environment

Figure 61 shows direct charging and battery electric vehicles are the most efficient with 73% of the energy going to motion. When regenerative braking is included, efficiency rises to about 90% of energy with less than 10% wasted. In contrast, a hydrogen fuel cell is only about 33% efficient and diesel about 20% and petrol only 16%. Some have them even lower.

Powering road vehicles with synthetic gasoline or diesel produced from green hydrogen would require 3.5 to 5 times as much renewable energy as a battery-electric vehicle (BEV) and should therefore not be allowed as a climate solution. [164] As we can see today, the solution is battery electric cars. The cost of the green hydrogen to make e-fuel makes e-fuels only economically viable for niche sectors, such as the replacement of Jet fuel. E-fuel is even less efficient than petrol. They may have zero emissions, but cost is much higher. Supplying just 10% of new cars with e-fuels instead of electrifying them will require 26% more renewable electricity generation in Europe. [165] Synthetic fuels should instead be prioritised for planes and ships, most of which cannot use batteries to decarbonise and which today burn fossil fuels that may be even more harmful to air quality.

You can understand the allure of low-emission fuels – you don't need the capital cost of a new vehicle to lower emissions. But when one assesses both the biofuel/e-fuel manufacturing process and the vehicle manufacturing process, EVs offer the lowest cost and fastest transition. The e-fuel is a costly distraction.

7.3.1 A Fuel for the Wealthy

There is vigorous lobbying by the fossil fuel industry for exemptions for e-fuels, but models show that, with higher e-fuel costs, running a car on e-fuels over five years will cost a driver $11,000 more than running a battery-electric car. High e-fuel costs will also make running second-hand cars on e-petrol around $11,000 more expensive over the same timeframe.

7.3.2 Nitrous Oxide Worse

While nitrogen oxides (NOx) are not themselves greenhouse gases, they are indirectly involved in the greenhouse effect and contribute to climate change. NOx emissions, primarily from fuel combustion, react in the atmosphere to form compounds such as ozone and nitrous oxide (N_2O), both of which are greenhouse gases. Cars powered by synthetic fuel emit as much poisonous nitrogen oxides (NOx) as fossil fuel engines. The emissions from a car running on petrol and three different e-petrol blends showed that the car running on e-petrol emitted toxic NOx at levels equal to those of standard E10 petrol (21 mg/km) and significantly higher levels of carbon monoxide and ammonia. [165]

7.3.3 CO_2 Emissions Worse

When burnt in petrol or diesel cars, synthetic fuels release similar amounts of CO_2 (and air pollution) as fossil fuels. Only by reducing GHG emissions in their production can they make themselves climate-neutral. This means the hydrogen used to produce these e-fuels must come from 100% additional renewable electricity. Additionally, the carbon molecules necessary to turn the hydrogen into the fuel should be captured from air. The previous CCS chapter shows CCS causes more emissions than it saves.

7.4 E-Fuel for Aeroplanes

So now we're in the same situation with jet fuel, and people ask "Well then what should we use in instead?" And I'm like, "Jet fuel, doh."

Jet aeroplanes use Jet-A1 fuel – a kerosene. Jet fuel packs a lot of energy for its weight and it is this energy density that has really enabled commercial flight. Today, there aren't any other viable options for transporting groups of people quickly over very long distances, so we're dependent on this type of fuel. The aviation industry is expected to double to over 8 billion passengers by 2050. Currently, most SAF fuels are made from cooking oil or waste and blended to about 50% of the final product. As we looked at above, SAF from the methanol pathway involves methanol synthesis, methanol-to-olefins (MTO) conversion, oligomerisation, and hydrogenation. Sources of bio-methanol are finite so it is not clear how SAF can continue past experimental stages.

7.4.1 Aviation Industry Global Emissions

The aviation industry's contribution to global emissions is small but meaningful, with few practical solutions. Commercial air travel is about 2.4% of all CO_2 emissions. A return flight between London and San Francisco has a carbon footprint of nearly 1 tonne of CO_2e per economy ticket. The share is expected to rise as demand for air travel increases. Beyond CO_2, aircraft also emit nitrogen oxides, water vapour, and particulates, which can have additional warming effects, estimated to double the climate impact of CO_2 alone. We discussed this in Part 1 of Optimistic Storm that the sector is hard to abate with current battery or hydrogen technology for long-haul flights. Short-haul is ripe for battery disruption, and flights under 2 hours are already likely candidates for battery-powered aeroplanes. Don't expect that to change for a decade as the industry is very conservative, risk averse and scarce of capital. The industry is increasingly turning to sustainable aviation fuels (SAFs) and e-fuels as a "drop-in" solution to reduce its carbon footprint.

7.4.2 Trials Using E-Fuels and SAF

Airlines and regulatory bodies worldwide are accelerating the adoption of SAF and e-fuels through mandates, partnerships, and trial flights. These fuels, chemically similar to conventional jet fuel, can be blended with traditional kerosene and used in existing aircraft and airport infrastructure without modification.

Mandates. The European Union's "ReFuelEU Aviation" initiative is a major driver, mandating a minimum share of SAF in jet fuel for flights departing from EU airports. This mandate begins at 2% in 2025 and is set to increase to 6% by 2030, with a specific sub-quota for synthetic fuels (e-kerosene) from 2030. The USA and UK have also set ambitious goals and are exploring mandates to spur production.

7.4.3 Airline and Industry Trials

Numerous airlines have conducted trial and commercial flights using SAF to demonstrate its viability and build a supply chain.

- United Airlines has been a notable leader, having made a significant investment in a SAF producer and committing to a long-term supply agreement.
- Lufthansa announced an "Environmental Cost Surcharge" to help offset the higher costs of SAF as it complies with new EU regulations.
- Japan Airlines completed a trial of a digital platform for trading the "environmental value" of SAF, a novel approach to financing its uptake.

- Wizz Air completed operational trials using SAF blends of 5% and 10% on flights from Barcelona and Charleroi, respectively.
- Air France-KLM is a partner in the "Take Kair" project in France, which aims to develop a domestic e-fuel industry for air transport.
- Swiss International Air Lines (SWISS) has been integrating solar-based SAF into its flight operations.
- Qantas has a target of 10% of its fuel use coming from SAF by 2030 and around 60% by 2050. [166]
- British Airways and Virgin Atlantic are among many other major carriers that have also conducted trial flights and secured long-term SAF supply agreements.

7.4.4 Cost SAF and Impact on Airfares

The primary obstacle to the widespread adoption of SAF is cost. The cost of SAF is currently two to seven times higher than conventional jet fuel, largely due to limited production capacity, high feedstock costs, and the technology's nascent stage. For instance, in 2022, the average price of SAF was estimated at around $2,400 per tonne ($1,000 to $5,000 depending on the pathway), while conventional jet fuel was $0.5 c/l or $680 per tonne. [167] This price disparity will inevitably be passed on to the consumer, leading to higher airfares. The exact increase is difficult to pinpoint and will depend on factors such as the blend percentage, the specific route, and the airline's cost structure. The current pricing suggests a carbon price of over $300 per tonne. One estimate suggests that a move to 10% SAF blending could result in a 3-12% premium on the ticket price.

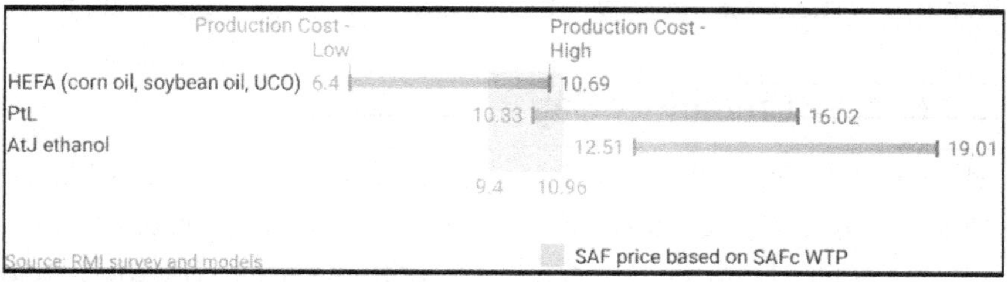

Figure 62 Production cost of SAF range and SAF price paid by airlines from RMI

In Figure 62, RMI shows that, assuming a 75% carbon intensity reduction, the SAFc WTP is a green premium of $2.34 to $3.93 per gallon, resulting in an SAF price range of $9.4 to $10.96 per gallon ($2,000 per tonne). With SAF production costs ranging from $6.40 to $19.01 per gallon, this green premium makes SAF cost-competitive with fossil-fuel-based jet fuel in some scenarios.

In response to the EU's SAF blending mandates, Lufthansa has already introduced an "Environmental Cost Surcharge" ranging from €1 to €72 per ticket, depending on the route. Other airlines are expected to follow suit. The dream is that as production scales up, technologies mature, and governments introduce supportive policies and incentives, the price gap between SAF and traditional jet fuel is expected to narrow over time.

7.4.5 Ethanol to Synthetic SAF

Bloomberg reported the opening of the Lanza-Jet plant, which received US government funds. They plan to make SAF (synthetic aircraft fuel) from biofuel produced using both traditional raw materials, including American-grown corn, and advanced technologies. The factory will produce 40 million litres (10 million gallons) of SAF and renewable diesel per year. Note. A 777-300 burns about 10,000 litres per hour or a typical flight from New York to London on the Boeing 777-300ER would then cost around $33,000 in fuel. That plant could fuel 4,000 hours of flying. There are 100,000 flights per day, many more of them more than 1 hour of flying!

Using all of the biofuel currently globally produced for aircraft could meet a quarter of jet fuel demand. Instead of using biofuel for other transport, it makes better economic and energy sense to use the bio-ethanol for jet fuel supply but there is a long way between 4,000 hours and 100,000 flights.

7.4.6 SAF Credibility Gap

There is a credibility gap here. It is a dream that is not playing out in real time. Much of this was based on a fall in hydrogen production costs, and as we saw in the Hydrogen Chapter, this is far from certain. The trials today generally use HEFA which is crop-based (corn, soybeans, grain) or waste-based (food waste, used cooking oils, residual wood). Those sources are limited. For full use of SAF we will need other production methods as discussed previously. SAF will need to be manufactured using hydrogen in the Power-to-Liquid (PtL) process. All the hype to date does not lower SAF and e-fuel costs, no matter how many media releases. Expect to see a switch over the next five years to other "sustainable" actions and we will see increases in ticket prices. All the evidence to date says making SAF a truly viable and scalable solution for the aviation industry's long-term sustainability goals has a long way to go.

7.5 Biofuel

Biofuels are a type of renewable energy derived from plant and animal biomass. They are categorised into different "generations" based on the type of feedstock used. First-generation biofuels use food crops, while second- and third-generation biofuels utilise non-food sources and advanced processes to be more sustainable. [1] Non-food includes crops such as corn, sugarcane, or soybeans, agricultural waste, or animal fats. The process typically involves fermentation of plant sugars or other biological processes. Ethanol is produced by fermenting the starches and sugars in plants.

Incredibly, we grow crops to burn in internal combustion engines in cars and trucks. It is one of the most bizarre energy failures of our society. The E10/D20 program and others similarly mandate petrol or diesel has 10%/20% ethanol. Primary ethanol sources are USA corn, Brazil sugar cane and Southeast Asia palm oil. The biofuel for petroleum started 20 years ago, when there were no real alternatives to fossil fuels. Wind and solar were in their infancy, and GM had just scrapped their EV.

Biofuel	Chemistry	Raw Materials	Use
Ethanol	Ethyl alcohol (C_2H_5OH)	Fermentation of sugar- or starch-rich crops like corn, sugarcane, wheat, and sorghum.	A "drop-in" fuel additive to gasoline (e.g., E10 or E85 blends) for cars and other road vehicles.
Biodiesel	Fatty Acid Methyl Ester (FAME)	Transesterification of vegetable oils (e.g., soy, canola, palm), animal fats, or recycled cooking oil.	A substitute for petroleum diesel in diesel engines, often used in blends (e.g., B20) for cars, trucks, and other heavy equipment.
Renewable Diesel	Hydrotreated Vegetable Oil (HVO)	Hydrotreating of vegetable oils and animal fats, often from waste sources.	Chemically identical to conventional diesel, it's a "drop-in" fuel for any diesel engine, including those in trucks, ships, and generators.
Biogas	Methane (CH_4) and CO_2	Anaerobic digestion of organic waste - animal manure, food scraps, sewage, and agricultural residues.	Used for heat and electricity generation, and can be upgraded to **biomethane** to replace natural gas in pipelines.
Sustainable Aviation Fuel (SAF)	Synthetic paraffinic kerosene (SPK)	Various feedstocks, including used cooking oil, animal fats, and non-food plants.	A key biofuel for the aviation industry, it can be blended with conventional jet fuel to reduce carbon emissions.

Figure 63 Biofuels - type, production and uses

Figure 63 shows some of the biofuels – either based on starch or oils. The drive to reduce oil use in transport led to ethanol mixed with petroleum and thought.

7.5.1 Ethanol E10 Fail

Whichever way you view the use of ethanol in petroleum, when you do the maths, it is an ecological and economic disaster. It is based on bad science, poor economics and corrupt politics. The process appears rational. Grow biomass, such as corn in the USA or sugar cane in Brazil, process it, and add it to petrol or diesel to reduce the amount of petroleum a country imports. It is a case of greenwashing from every player. Michael Grunwald in his recent book We Are Eating the Earth: The Race to Fix Our Food System and Save Our Climate [168] points out that a big study was done at the Argonne National Laboratory that concluded ethanol was maybe 20% better for the climate than gasoline. When you dig deeper, it is really weird.

- Most of the study reviewed how inefficient ethanol was.
- Ethanol is produced from corn.
- Farmers have to till the land with tractors.
- Fertilise the corn.
- Operate tractors for harvesting the corn.
- Ferment and brew the ethanol.

Ethanol produces much more greenhouse gas emissions than using gasoline alone. Argonne had completely ignored land use.

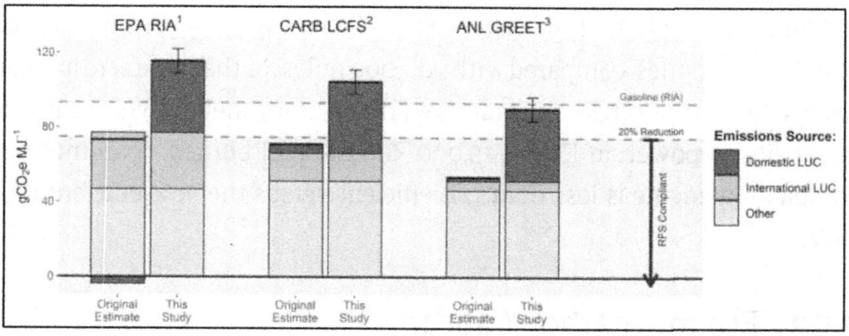

Figure 64 Emissions from growing corn for ethanol vs gasoline

Figure 64 shows how emissions from corn are much higher due to land use impacts. The inputs create a fuel that is more carbon-intensive than just burning petroleum. For example, with gasoline, you burn it in a car, and the carbon goes up into the sky. Corn, you burn it in the car and the carbon goes up into the sky, but then you grow corn in the field and that absorbs carbon from the sky.

Yet vast subsidies to keep farmers on the land are doled out each year, and using E10 fuel is higher emissions.

- Causes 24% more global warming.
- Raises the price of corn by 30%.
- Increased fertiliser use by 3-8%.
- Degrades water by 3-5%.

So while it is clear that ethanol mixtures in gasoline are a bad policy, ethanol continues to be used in transportation due to agriculture's strong influence on energy policy. [169] It gets worse.

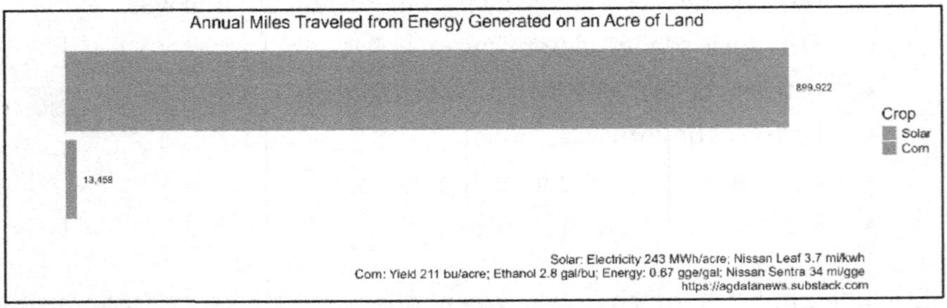

Figure 65 Distance travelled on ethanol from corn vs solar panels

In Figure 65, we see that the distance travelled using ethanol from 1 acre of corn is 14,500 miles, compared with 900,000 miles for the same acre in solar panels. Just ½ of a single-car garage's worth of solar panels provides enough electricity to power an EV for 15,000 km. Ethanol burned in an internal combustion engine is less than 17% efficient versus the 90% efficient in a BEV.

7.5.2 Ethanol Carbon Capture

A proposal in the USA to use carbon capture to produce ethanol from corn has to be one of the most stupid emissions-reduction initiatives in a long time! The \$5.6 billion project proposes to add CO_2 capture to yeast fermentation process. USA and Brazil farmers are the largest group to grow corn to ethanol. The plan is to build pipelines under the properties of hundreds to thousands of landowners across multiple states to transfer the CO_2 to an underground storage facility in the Dakota "Badlands".

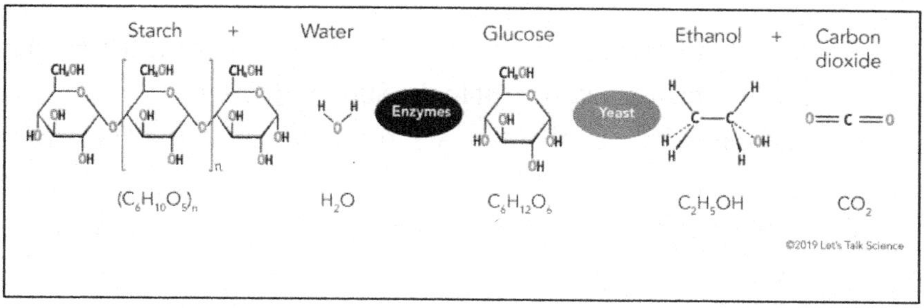

Figure 66 Process of turning starch to ethanol and CO_2

Figure 66 shows the process to turn starch to ethanol, with the waste product of CO_2.

- Navigator is proposing a 1300-mile pipeline.
- Wolf - a 380-mile pipeline.
- Summit - a 600 km pipeline, with 34 ethanol plants at a cost of \$5.6 billion. CO_2 captured is expected to be about 9.5 million tonnes.

In theory, capturing CO_2 from the fermentation process during ethanol production may reduce ethanol's overall lifecycle CO_2 emissions by~30gCO_2/MJ. More likely, the CO_2 would be used to extract more gas and oil from older gas fields.

Jacobson used a complex model [170] comparing building an ethanol plant and a CCS plant versus building wind farms needed to supply the energy for the ethanol.

- Production of E85 from ethanol with carbon capture, followed by E85's use in a 2023 Ford F-150 four-wheel drive (4WD), 8-cylinder ICE pickup.
- Production of wind electricity, followed by its use in a 2023 Ford - 150 Lightning 4WD extended range BEV.

The model also assessed 2 wind scenarios across the the five states; Iowa, Nebraska, South Dakota, Minnesota, and North Dakota. Scenario A was based on coal-fired electricity, and Scenario B assumed these states went all-wind.

Needless to say, the economics did not stack up.

- A 12–20 times better return to have wind and electric trucks than the $ 5.6 billion investment in the Summit project.
- Emissions from wind were 25% to 60% less than the CO_2 carbon capture proposal.
- Fuel savings was $66.9–$111 billion (USD 2023) over 30 years for the BEV
- Higher upfront costs with BEV than with ICE vehicles but that higher cost was offset within two years.
- Cost of deploying wind – enough for the vehicles or for all of the states.

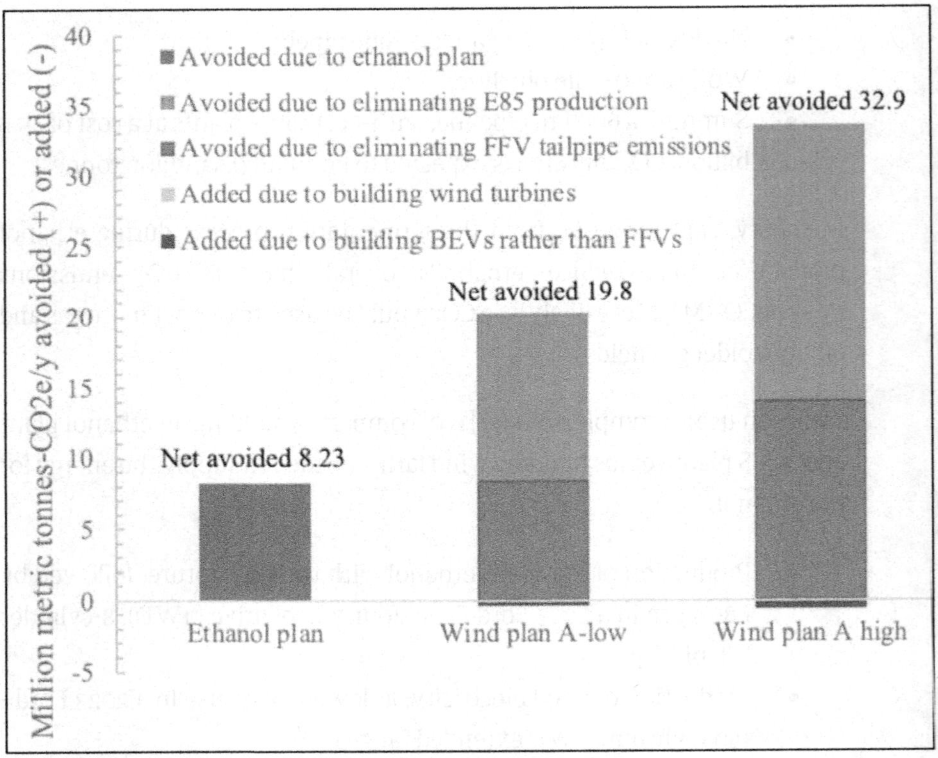

Figure 67 Carbon capture versus investment in wind and electric vehicles avoided CO_2

In Figure 67 we see that the carbon emissions of using wind versus carbon capture in ethanol production are much less and much less expensive. The carbon capture adds over 8 million tonnes of CO_2 emissions due to the energy for capture, processing, transport and burial. Stopping growing corn for E85 avoids another 11 million tonnes, and moving to electric vehicles and using wind avoids over twice the emissions. The challenge is one of politics. Technically and economically, going electric makes sense. It is just politics. Farmers and businesses are reliant on subsidies. Governments spent money to keep those farmers voting. Utter stupidity.

7.5.3 Ethanol Other Uses

Ethanol for transport is not only inefficient, but it also contributes to climate change through broad-acre agriculture. There are other uses. A potentially significant future use as it is a key building block for a range of bio-based materials, offering a renewable alternative to petroleum-based feedstocks. Ethanol is widely used in pharmaceuticals and personal care products. In medicine, it serves as a solvent to dissolve active ingredients in oral medications, allowing for easier absorption by the body. It acts as a preservative, preventing the growth of bacteria and fungi and extending the shelf life of liquid medications. It is a common antiseptic and disinfectant used in hand sanitisers and medical wipes, and for sterilising medical equipment and surfaces. In the food and beverage industry and in chemical and industrial industries, it is a powerful solvent, dissolving many organic compounds in a myriad of uses, whether food products, paints, adhesives, inks, to name just a few.

We discussed plastic in Parts 1 and 2 and its ever-increasing pollution of our planet, environment, bodies, and brains, and looked at the implications for our health. Ethanol's role in the plastics sector could extend beyond being a "green" ingredient; it's a fundamental chemical that can be converted into a variety of high-performance plastics and other materials. The primary way ethanol is used to produce bio-based materials is by converting it into ethylene. The ethanol dehydration process is a critical first step. Once ethanol is converted to bio-ethylene, it can be used to produce a variety of bioplastics that are chemically identical to their petroleum-derived counterparts, including:

- Bio-based Polyethylene (Bio-PET): This is one of the most widely used plastics in the world, found in everything from food packaging and bottles to plastic bags and consumer goods. The production of Bio-PET from sugarcane ethanol, for example, allows for a material with a lower carbon footprint that can be integrated into existing recycling streams. [171]
- Bio-based Polypropylene (Bio-PP) polymer is used in a wide range of applications, including automotive parts, packaging, and textiles. By producing it from bio-ethanol, manufacturers can reduce their reliance on fossil fuels while still creating a durable, versatile material. [172]
- Bio-based Polyvinyl Chloride (Bio-PVC) is a petroleum-based plastic, it too can be made from ethanol derived from renewable sources. Bio-based versions offer comparable durability and performance characteristics, making them suitable for a variety of applications, from pipes and cables to flooring and siding. A marketing report predicts a growth of 20% CAGR from a base of $1 billion. [173] It has a long way to go – the global fossil fuel PVC market is a $86 billion market.

Beyond the creation of these common plastics, ethanol also serves as a feedstock for the production of other bio-based materials and chemicals:

- Polyhydroxyalkanoates (PHAs): Ethanol can be used as a carbon source for microorganisms that produce PHAs. These are a class of biodegradable and biocompatible polyesters that have a wide range of uses, including medical implants, packaging films, and agricultural products. [174]
- Butadiene. Traditionally, butadiene is produced from petroleum but ethanol can also be a feedstock. Butadiene is a key building block for synthetic rubber and various plastics. Ethanol could up new possibilities for sustainable manufacturing. [175]
- Plasticiser or Blending Agent. Ethanol can be used as an additive in the production of other bioplastics, such as Polylactic Acid (PLA), to modify their properties. By blending ethanol-derived materials with other biopolymers, manufacturers can enhance characteristics like flexibility, heat resistance, and tensile strength,

creating customized materials for specific applications. [176]

So we can see ethanol can be a feedstock for bio-based materials. We should not be burning it. Current research shows it can help build a more sustainable, circular economy without using fossil fuels. It could provide a way to decouple plastic production from fossil fuels and allow for the creation of new materials with superior or more targeted properties. We need to go back and look at disruption to see whether ethanol and other bio solutions can compete with the dominant supply of chemicals from the fossil fuel industry. Expect fossil fuel plastic producers to try to outcompete and prevent a loss of market share. They are going to lose the transport sector, which is 65% of their market. They are fighting tooth and nail now. What would they do if their plastics market was at risk as well?

7.5.4 Biofuel By Country

The largest fuel ethanol producers are the USA at 61,000 million litres, Brazil at 33,000 million litres, the EU at 5,500 million litres, India and China, and then the "Rest of World" data. [177] About 0.36 million tonnes of Australian wheat is used in ethanol manufacture along with 122 million litres of molasses in 2022. (See report Ethanol Growers) That wheat also needs diesel for crop production, water, pesticides and takes over 200,000 hectares for the actual wheat production. Average wheat yields in Australia is about 2.8 tonnes per ha. And it just goes up in smoke, utilised at less than 18% in combustion engines.

7.6 Conclusion: E-Fuel & Biofuels

Both synthetic ethanol and synthetic methanol could be sustainable fuel solutions. Differences in feedstock, production process, and overall carbon footprint will drive adoption. The hydrogen to PtL (power to liquid) making E-methanol with DAC appears to offer the most promising long-term solution for achieving significantly lower emissions, yet both the hydrogen and the DAC are very energy-intensive processes. Current pricing is at least five to ten times more than petrol and diesel. Continued research and development are crucial to improving their efficiency, expanding production capacity, and ensuring their role in a sustainable future. The primary challenge to sensible decision-making is that engine-makers and oil companies want a loophole for e-fuels so they can continue selling combustion engines, and oil companies want to delay the move to electrification.

E-fuel and biofuels should not be used for any land transportation or where batteries can power transport. Those producers are similar to hydrogen producers who do not want electrification. The producers of expensive and polluting fuels are looking to push their products and make consumers pay a green tax.

Fundamentally, we need to stop burning stuff. Harvesting electrons and electrifying everything is more efficient than burning molecules for their energy. The use of E-fuel and biofuel is merely a transition strategy that allows existing engines designed to run on oil to use a low-carbon source of fuel. This transition from burning stuff to using harvested electrons is a very messy period with a large number of vested interests. The 1.4 billion cars have an average age of about 11 years, so within 20 years, there will only be battery electric cars and the fight will be over. However, the 29,000 passenger aircraft globally fly for a median of nearly 20 years, so it will take decades to design new planes and replace the existing fleet with more alternative-fuelled aircraft. In the meantime, we will continue to see airlines attempt to reduce emissions, and passengers will pay.

Conclusion of Biofuel and E-Fuel:

- Time to stop burning plants.
- Solar to electricity is orders of magnitude more efficient and lower cost for transport.
- Changes to the electrification of land transport will eliminate most use for biofuels and e-fuels.
- E-fuel is based on the hydrogen myths and needs to be stopped except for a restricted use set in something like flying or long distance ocean freight.

8 Communications

Explore Communications:

- Why Have a chapter on Communications?
- Role of communications in the new energy and AI world?

Sometime during the last ten years, the stars aligned in SpaceX's favour. Once a small start-up with a reputation for blowing up its rockets, the company has ballooned into spaceflight's undisputed kingpin – largely thanks to its cost-effective, partially reusable Falcon 9 rockets. Along with this, developmental delays and even political shunning brought once-mighty competitor launchers to their knees, leaving the market for the taking from SpaceX competitors. SpaceX now accounts for 80% of the payload mass launched into orbit. Its estimated market capitalisation in 2024 was about $400 billion, and it plans to IPO in mid-2026 at a valuation of over $1.4 trillion. Questions are being asked about why it has paid no tax yet has made over $15 billion in revenue and billions in profits over the past couple of years, and has received government contracts worth over $38 billion since 2003. The Washington Post estimates that another $11.8 billion will come in the next few years, thanks to 52 active contracts. SpaceX may not pay any tax for a decade.[178] Its contracts now include:

- NASA crew transportation to ISS.
- Lunar Lander for NASA Artemis program.
- Resupplying the ISS.
- NASA vehicle to deorbit the ISS.

- Launch services for NASA.
- Launch services for US Space Force.
- Starlink Government Contracts.

There are other contracts that are government-secret. While there are plenty of knockers, it is a tribute to SpaceX and its founder, Elon Musk, that they made SpaceX work. In particular, Starlink is a service that can bring communication to every place on earth.

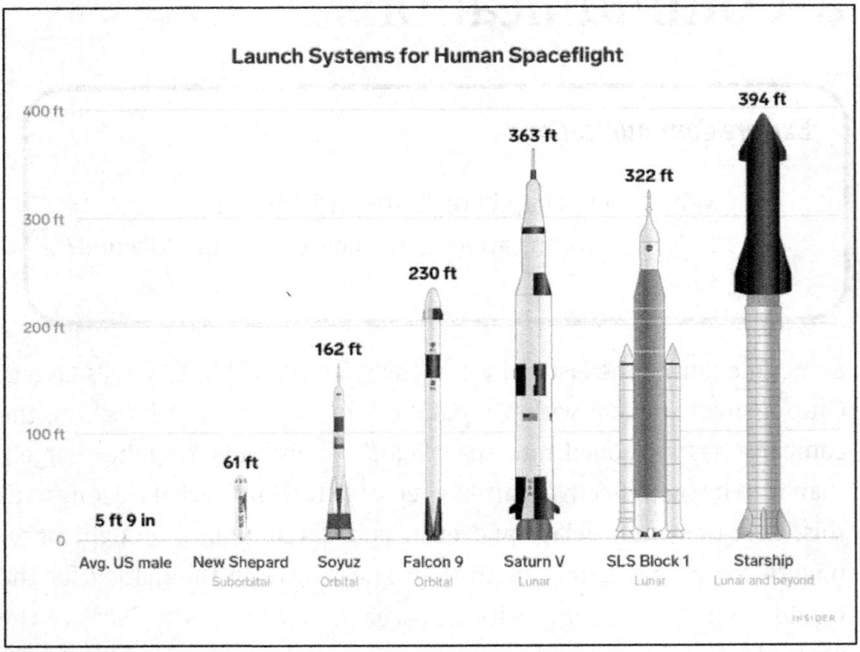

Figure 68 Size of rockets for human space flight (from Business Insider)

Figure 68 shows rockets have got much bigger, with payloads much bigger. Falcon 9 has enabled Starlink to get stuff into space. Lots of it. The cost of lifting payloads into space plummeted with the Falcon 9, with SpaceX banking on Starship to further reduce costs.

- Space Shuttle = $52,000
- Saturn 5 = $ $6,000
- Soyuz = $7,000/kg
- Falcon 9 = $2,000/kg
- Starship = $200/kg

The cost of putting a kilo into space is 10 times lower than it was a decade ago. Some even suggest costs of Starship will fall to $30/kg enabling the establishment of a colony on the Moon or Mars. The latest thought bubble from Musk is that they will put AI data centres in space beginning in early 2030. They need their Starship to work; Tesla has purchased solar panel-making equipment for making over 100 GW of panels per year; they have designed AI chips for these satellites; and they have light-speed communications.

8.1 Starlink

We will explore satellite communication as it will be a technology used in this era four in areas such as transport, energy, robotics, and health. A Starlink satellite from SpaceX [179] has a lifespan of approximately five years and SpaceX has approval from FCC to have as many as 42,000 satellites in this so-called mega constellation. The current V2 Starlink satellite version weighs approximately 800 kg at launch, almost three times heavier than the previous generation satellites at 260 kg. The low Earth orbit (550 km LEO) satellite had 5,874 in orbit, servicing over 6.5 million subscribers (Jul 2025). By mid 2026 over 10,300 are in operation and is churning them up about 70 per week. Starlink requires ground stations so approval is country by country - currently all 7 continents, and in over 150 countries. Starlink brings high-speed internet to people with little or no access. Education is as close to a search or AI query as is entertainment.

Starlink sells direct to consumers and businesses but has partnered with existing telecommunications companies. The introduction of satellite phone communication will further disrupt the existing communication system.

8.2 Conventional Satellites

Some of the largest legacy operators in geostationary orbit have made moves over the last decade to respond to the upcoming competition from SpaceX. The only operational low-Earth orbit internet constellation besides Starlink was launched by OneWeb, which primarily sells capacity to existing internet providers, who then distribute services to individual consumers, in contrast to Starlink providing services directly to homes and businesses.

Eutelsat, the third-largest operator of geostationary satellites, merged with OneWeb last year, creating a company with a blended offering of GEO and LEO services. Viasat, a pioneer in satellite internet services using dedicated spacecraft in geostationary orbit, last year purchased Inmarsat, which specialised in providing connectivity to aeroplanes and ships. Recently, SES acquired Intelsat and stands apart due to the size of the satellite fleets. Founded in 1985, SES currently operates 43 geostationary satellites, plus 26 broadband spacecraft in medium-Earth orbit (MEO) a few thousand miles above Earth. These MEO satellites operate in a kind of middle ground between LEO and GEO satellites, offering lower latency than geostationary networks, while still flying high enough not to require hundreds or thousands of spacecraft to blanket the globe.

Intelsat has 57 geostationary satellites, primarily for television and video relay services. They offer coverage over 99 percent of the world, and provide services through a range of communication bands. For now, LEO broadband satellites in the Starlink and OneWeb networks beam signals to user terminals in Ku-band, but the new combined network will span Ka-band, Ku-band, X-band, C-band, UHF, and secure bands tailored for military use. Internet signals coming from a GEO satellite, like a Viasat spacecraft, typically have a latency of about 600 milliseconds. SES's O3b network in medium-Earth orbit provides signals with a latency of about 120 milliseconds, and Starlink latency ranges between 25 and 60 milliseconds.

In conclusion, a space race is still underway for three purposes.

- Providing internet services to businesses and consumers globally. Satellites cannot compete with fibre in cities with millions of access points for 4G/5G/6G services but can provide fast internet services for all the areas outside of the telecommunication networks. We can expect to connect anywhere in the world within a few years.
- Preparation to go back to the moon, and eventually set up a colony on Mars. Musk rationalises this by saying it would provide a backup plan for life on Earth in the event of a terminal event, such

as a meteor impact or nuclear war, and he is focused on unmanned flights by 2026 and crewed flights by 2030.

- Provide long-distance travel in sub-orbital flight with a Sydney to London flight in 2 hours, not 22, and there are various people who say this will be available by the mid-2030s.

Space flight is not needed to end the fossil fuel era, but imagination can run wild about the advances possible in this 4th era of transportation.

8.3 Other Rocketmen

Besides SpaceX, several other companies are actively involved in the rocket and space industry. Some prominent Western examples include Blue Origin, Rocket Lab, Relativity Space, United Launch Alliance (ULA), Virgin Galactic, Firefly Aerospace, and Arianespace. [180] These companies offer a range of services, from orbital launches to space tourism and satellite deployment. These include:

- Jeff Bezos founded Blue Origin, which is developing New Glenn, a heavy-lift orbital launch vehicle, and New Shepard, a suborbital vehicle for space tourism.
- Rocket Lab specialises in small satellite launches with its Electron rocket and is developing the Neutron rocket for larger payloads.
- Relativity Space focuses on 3D-printed rockets, with its Terran 1 (now retired) and Terran R (in development).
- United Launch Alliance (ULA) is a joint venture between Boeing and Lockheed Martin. ULA provides launch services for government and commercial clients, using rockets like Atlas V and Vulcan Centaur.
- Virgin Galactic is primarily focused on space tourism, Virgin Galactic offers suborbital spaceflights for passengers.
- Firefly Aerospace is developing launch vehicles like Alpha and Beta, as well as lunar landers.
- Arianespace from Europe provides launch services for satellites using various rockets, including those from ArianeGroup.

8.4 China Space Activities

The West tends to focus on SpaceX and its disruption. Russia and Europe are making minimal progress in space. However, China's space program has achieved remarkable advancements in recent years, establishing itself as a major player in space exploration and technology. [181] The West does not hear about them or recognise that they are probably as advanced as the USA. Notable achievements include lunar and Martian missions, the development of a space station, and significant progress in satellite technology and launch capabilities. Features of the dragon in flight are:

- China has one of the most active space programs in the world. With space launch capability provided by the Long March rocket family and four spaceports (Jiuquan, Taiyuan, Xichang, Wenchang) within its border, China conducts either the highest or the second highest number of orbital launches each year.
- Lunar exploration such as Chang'e-5 mission successfully returned samples from the Moon's surface, and the landing of Chang'e on the far side of the Moon.
- Tianwen-1 mission landed on Mars with the Zhurong rover.
- Tiangong Space Station is Chinas own space station. Tiangong, provides long-duration manned spaceflight and in-orbit construction.
- They have made considerable progress in satellite technology, including the development of advanced communications, navigation, and remote sensing satellites including their own GPS network.
- They are addressing the challenge of space debris, actively participating in discussions on international rules and mechanisms for ensuring the long-term sustainability of outer space activities.
- Their commercial space sector is also rapidly developing, with companies like Galaxy Space and SpaceTy launching satellites developing innovative technologies.
- With space life science research they investigate the effects of space on living organisms and are developing technologies for long-term space exploration.

- China is doing some international cooperation and actively participating in discussions on outer space governance and working with other countries on various space-related projects.
- China plans to send astronauts to the Moon and establish a lunar research station.
- They are planning future missions to explore other planets and celestial bodies beyond the Moon and Mars.
- They are building out space infrastructure, including advanced launch vehicles, space stations, and satellite constellations.

China's space program is increasing in capabilities. The CCP has ambitions to be a major player in the space landscape. The military space race will be going as fast or faster than the civilian ramp. If they can cooperate with other nations, then the world will benefit.

8.5 Undersea Cables

The first transatlantic cable in 1858 demonstrated the potential for rapid communication between continents. This revolutionised business and governance, and today, there are approximately 1.4 million kilometres of undersea data cables globally. This vast network of cables carries the vast majority of international internet traffic, acting as the backbone for global communication. A small number of organisations privately own about 99% of cables. About 1% of submarine cables are owned or partially owned by a government entity.[182] They are not cheap and cost upwards of $30,000 /km, meaning a longer cable can easily run in the hundreds of millions of dollars. Recent trans-oceanic cable bids have reached upwards of $250-$300 million for trans-Atlantic and $300-$400 million for trans-Pacific.

8.6 Conclusion: Communications

The world relies on very few companies for communication. Space communication has revolutionised data communication. Starlink is now providing standard mobile phone connectivity – albeit slow due to the size of the phone aerial. We have seen data speeds increase from 300 bps in the 1970s to over 224 Tbps. That is over 700 million times faster over 50 years. Experimental fibre is now a thousand times faster at 29Pbps. The World

Bank said 2014 [183]

"a 10% increase in broadband penetration results in a 1.38% increase in gross domestic product (GDP) growth in low and middle-income countries"

and with the rise in AI we will continue to see demand for communications continue to rise. Starlink and others will bring the Global South to full connection over the next few years. We will continue to see the expansion of fibre on the ground and across oceans. It is unlikely that a satellite will carry the bulk of the data, but satellites now can provide reasonable data communication within days, which used to take years. With over 7.43 smart phones globally and over 15 billion smart devices today, no one or nothing will be without global communication by 2030.

Conclusion of Communications:

- Better communication infrastructure is needed for the transition in renewable energy and AI
- The Global South needs communication as much as the developed north.
- All Governments, both autocratic and democratic will want to restrict increasing and global reach

9 Conclusion: Part 3

Conclusion of Part 3: Technology Fails

- Change is in full swing as we harvest electrons for our energy needs.
- There's an Optimistic Storm in energy! But also in AI, health, transport and labour.
- There is little optimism for continuing to burn molecules.
- Hydrogen, nuclear, CCS, or biofuel or e-fuel are all part of the last era of energy. Fossil fuel companies will try every last trick to slow down the transition, even to the extent of ignoring physics, technology and economics.

When I first drafted Optimistic Storm, it expanded to over 120,000 words so I decided to break it into three parts. As discussed in Part 1, we are in the 4th Era of energy and transport and we can see the end of the fossil fuel era. Humans used grass, grains, and animals, relying on their muscles for transport for nearly a million years in Era 1. It was only when they domesticated animals in 6,800 BC that we began to use the muscles of beasts of burden. Less than 300 years ago, with the invention of the steam engine, we replaced animal power with steam power and so moved into Era 2. It took a century to develop steam engines for trains and ships and those technologies spawned the Industrial Revolution and global colonisation with steam trains and steam ships. Era 3 driven by oil and gas, took over 150 years to go from initial use to widespread application. We got cars in 1900, and aeroplanes soon after. Today, after 70 years of harvesting electrons, we are now firmly in Era 4 - a disruptive phase change—moving away from burning molecules for energy to harvesting electrons using solar, wind, and batteries. Some haven't recognised the magnitude of this shift and cling to the idea of burning fuel for energy. The consequences of the end of the fossil fuel era will see massive geopolitical consequences. Twenty-four countries will no longer lord it over those 150 countries through control of the fossil fuel supply. Those 24 companies that create 71% of the carbon and methane emissions are stranded assets, and unless they change, they will be out of business within two decades. We see their all-out efforts to distort, lie, cheat and try to block the inevitable change to clean energy and transport.

In Part 2 of Optimistic Storm, we explored how AI was disrupting food, work, health and agriculture. AI will radically change our lives -the tasks that knowledge workers do will be augmented or AI will replace their jobs completely. With blue-collar workers, AI will replace them completely in conjunction with humanoid robots. We are much earlier in the phase change transition, and there are still many unknowns, but it is playing out before our eyes.

In Part 3 of Optimistic Storm, we have reviewed technologies including nuclear, hydrogen, CCS, DACs, E-fuels and biofuels in the broader context from changing from molecular energy to electrons. Disruption is complex, and people make mistakes and try to pick winners. But all these technologies are molecular, and nuclear is one such fuel people have trouble saying goodbye to. No one likes stranded assets. Despite being three times as costly, with no effective waste program and with an ongoing threat of military use and accidents, we have seemingly intelligent people who do not seem to understand that the world has changed. They either do not recognise that change or they wilfully ignore it. All the subsidies and efforts lavished to breathe new life into a dying technology does nothing more than slow down the change from burning rocks to harvesting and storing electrons. Wind, solar, and batteries are the economic future and will bring over 800 million people out of energy poverty and provide abundance to all. If only we could get politicians to assist, not hinder.

Many politicians and well-meaning financiers thought hydrogen to be part of this new Era 4. The reality is physics does not lie. Part of this push for hydrogen was for transport, but electric battery-powered transport is four times as energy-efficient. We see opposition from auto manufacturers who want to have an emissions-free fuel for their existing internal combustion engines. Sadly, they exhibit behaviour incumbents often show – change has to come from outside their industry. Most have the faster horse fallacy and don't see that the industry they spent their lives in is changing. They dismiss the impact of technology from outside their industry, for after all, who would have thought the portable music player industry would disrupt the vehicle industry? Or that the vehicle industry would disrupt the electricity grid industry. Or that humanoid robots and AI will spell the end of human toil.

We can be very optimistic. In this 3-part series, we explored how real climate change will be catastrophic for billions of people. We need to stop emissions as fast as possible to minimise the damage. We have the technologies to stop emissions, but we don't have technologies to prevent floods, storms, sea rises, lethal humidity deaths, and to prevent the hundreds of billions people who become climate refugees.

Let's be optimistic that we can stop digging the hole we have dug with emissions and the associated climate change. Stop investing in those things that slow it down. Only then do we have to focus on how we can remediate the damage done so that our children and grandchildren can have a world at lower risk than we have.

Final Conclusion:

- We have to stop polluting the earth.
- Have need to stop burning molecules for energy.
- Harvesting electrons for electrical energy is the cheapest, fastest and most reliable method.
- Fossil fuel companies and governments continue to slow down progress by giving credence to technologies that are on their death spirals.
- By understanding the technology, the economics and the politics, we can refute the politicians. We can change them from slowing down the phase change disruption to speeding it up.
- Above all – we must remain optimistic.

10 Dedication

To my long-suffering partner Judy, as I talk about incessantly. To those colleagues and friends who have worked with me over decades. To my children and grandchildren, for whom I wish to leave the world a more equitable place, a place where we live in harmony with nature.

11 Limitations on Information

Change may be rapid. Information in this book may be dated. Steinbeck in 1929. "How did you go bankrupt? Slowly and then suddenly." Innovation can take years to go from lab to full commercial deployment. Globally, R&D is over $2.2 trillion, and ranges from minimal to 6% in Israel.[184] I've been exposed to thousands of startups in my career, and it's this tsunami of great ideas. Many fail to see the light of day, while others change the world. A few get media, often due to companies needing money but most are quietly introduced. That's why I believe there is a storm of optimism that will change our society. Most need some regulatory approval or construction so from idea to product may be 10 or 20 years - especially in health care. Some appear almost instant. ChatGPT AI was released in 2022, and became the fastest deployed software, but the birth of AI was in the 1950's!

Hang onto your hats. There are thousands of good things happening in clean energy, transportation, Artificial Intelligence, robotics, food and health care and will flood our lives over the next 2 decades. Understand that it is seldom just one breakthrough. The combination of multiple technologies gives change bigger than just the sum - it's multiplicative.

12Glossary

We use TLA's - three letter acronyms (WTF) or even FLAs (four letter ones), so here are some TLAs, FLAs and phrases used throughout this book.

M = million
Billion = billion or 1,000 times more than a million
T= 1,000 billion
J=joule. A measure of energy. A joule of energy applied for 1 second is a watt. Therefore, a Wh is 3,600 joules.
W= watt. A more commonly used form of energy is measured on a day-to-day basis. An incandescent light bulb might be 60W. A LED 7W would produce about the same lumens.
Wh = watt hour. How much energy for an hour. A 60W light bulb would use about 60Wh if on for 1 hour. That energy is converted into light energy, and heat energy.
kW = 1,000 Watts
MW = 1 million Watts.
Sun = a huge free nuclear fusion and fission energy source, that has lasted 6 billion years and due to last another 6 billion years
EEI = Earth energy imbalance. The Earth (more accurately, the ocean) receives about 1370 watts per square meter (W/m^2) in the tropics. All but 1 to 2 is irradiated back out.

RE = Renewable Energy
ICE = internal combustion engine
EV = Electric vehicle
BEV = Battery Electric Vehicle
Electron excitation occurs when silicon is "excited" by solar energy (light), releasing a series of electrons with higher energy. If captured, we have how a solar panel works.
Hybrid = an electric vehicle with an internal combustion engine
PHEV = A hybrid that can be plugged in to charge the battery
FCEV = a hybrid vehicle with an electric motor using a hydrogen fuel cell to charge the battery.
IPCC = Intergovernmental Panel on Climate Change
Molecular energy. Energy is released from a chemical reaction from one compound to another. For example, oxidising (burning) a hexane molecule to CO_2 and water releases energy.
Moore's Law = the number of transistors on a chip doubles every 2 years (1965).
Swanson's Law = states solar module prices drop 20% for every doubling of installed capacity. Costs go down 75% per decade.
Wrights Law = As production increases, costs decline at a consistent rate as the cumulative (Wright 1936), sometimes referred to as learning curve effect. Predicts learning rate of 20-25% .
A climate denier = someone who ignores data and denies climate change.
A climate delayer = someone who accepts climate change but takes not action.
A nuclear advocate = A Climate delayer who proposes more nuclear in 15 years and keep burning fossil fuel.
A hydrogen advocate = a climate delayer with hopium and little or zero science understanding.
Malthusianism. A 19th century myth that there is a population bomb and thinks eugenics is a solution, as Germany did in the 1930's, or supports the UN Fund for Population Activities (UNFPA).
Doomster = sees climate change and thinks we cannot change the progression of climate change impacts and escapes to Bali to do regenerative agriculture.
LCOE = LCOE, or Levelized Cost of Energy, is a metric that calculates the

average total cost to build, operate, and maintain an energy generation facility over its lifetime, expressed as a cost per unit of energy produced (for example dollars per megawatt-hour or cents per kilowatt-hour). It serves as a standardised way to compare the economic viability of different energy technologies,

13 Table of Figures

14 Citations

[1] A. Bilal and D. R. Känzig, 'The Macroeconomic Impact of Climate Change: Global vs. Local Temperature', May 2024, *National Bureau of Economic Research*: 32450. doi: 10.3386/w32450.

[2] 'Climate crisis costs the world 12% in GDP for every 1°C temperature rise, and other nature and climate stories you need to read this week', World Economic Forum. Accessed: Sep. 17, 2024. [Online]. Available: https://www.weforum.org/agenda/2024/06/nature-climate-news-global-warming-hurricanes/

[3] 'Amazon.com: Brighter: Optimism, Progress, and the Future of Environmentalism eBook?: Dorr, Adam: Kindle Store'. Accessed: Aug. 15, 2024. [Online]. Available: https://www.amazon.com/Brighter-Optimism-Progress-Future-Environmentalism-ebook/dp/B0BNYC1GWY

[4] 'Norway's Equinor forced to withdraw key carbon capture claim | RenewEconomy'. Accessed: Jan. 20, 2025. [Online]. Available: https://reneweconomy.com.au/norways-equinor-forced-to-withdraw-key-carbon-capture-claim/

[5] 'Networks of climate obstruction: Discourses of denial and delay in US fossil energy, plastic, and agrichemical industries | PLOS Climate'. Accessed: Jan. 20, 2025. [Online]. Available: https://journals.plos.org/climate/article?id=10.1371/journal.pclm.0000370

[6] 'The Electrotech Revolution', Ember. Accessed: Sep. 21, 2025. [Online]. Available: https://ember-energy.org/latest-insights/the-electrotech-revolution

[7] 'Vaclav Smil - Wikipedia'. Accessed: Apr. 26, 2024. [Online]. Available: https://en.wikipedia.org/wiki/Vaclav_Smil

[8] IChemE, 'Uses of Hydrogen in Industry'. Accessed: Jul. 23, 2024. [Online]. Available: https://www.thechemicalengineer.com/features/uses-of-hydrogen-in-industry/

[9] B. Soergel *et al.*, 'A sustainable development pathway for climate action within the UN 2030 Agenda', *Nat. Clim. Change*, vol. 11, no. 8, pp. 656–664, Aug. 2021, doi: 10.1038/s41558-021-01098-3.

[10] D. Castelvecchi, 'How the hydrogen revolution can help save the planet – and how it can't', *Nature*, vol. 611, no. 7936, pp. 440–443, Nov. 2022, doi: 10.1038/d41586-022-03699-0.

[11] A. and P. R. Center (APRC), 'The Earth's core contains large amounts of hydrogen, held 50 times more water in the past than in the oceans

of today, research group including the University of Tokyo proves', ScienceJapan. Accessed: Jun. 12, 2024. [Online]. Available: https://sj.jst.go.jp/stories/2021/s0906-01p.html
[12] D. Slanger, 'Hydrogen Reality Check: All "Clean Hydrogen" Is Not Equally Clean', RMI. Accessed: Aug. 21, 2025. [Online]. Available: https://rmi.org/all-clean-hydrogen-is-not-equally-clean/
[13] 'Hydrogen Shot', Energy.gov. Accessed: May 17, 2024. [Online]. Available: https://www.energy.gov/eere/fuelcells/hydrogen-shot
[14] C. B. Staff, 'Guest post: Why hydrogen cars are being outsold by Ferraris', Carbon Brief. Accessed: May 17, 2026. [Online]. Available: https://www.carbonbrief.org/guest-post-why-hydrogen-cars-are-being-outsold-by-ferraris/
[15] '(PDF) Green hydrogen production ensemble forecasting based on hybrid dynamic optimization algorithm'. Accessed: Aug. 06, 2025. [Online]. Available: https://www.researchgate.net/publication/371948969_Green_hydrogen_production_ensemble_forecasting_based_on_hybrid_dynamic_optimization_algorithm?_tp=eyJjb250ZXhoIjp7ImZpcnNoUGFnZSI6Il9kaXJlY3QiLCJwYWdlIjoiX2RpcmVjdCJ9fQ
[16] 'Natural Hydrogen: A Potential Clean Energy Source Beneath Our Feet', Yale E360. Accessed: Aug. 10, 2025. [Online]. Available: https://e360.yale.edu/features/natural-geologic-hydrogen-climate-change
[17] 'Natural hydrogen: sources, systems and exploration plays | Geoenergy'. Accessed: Aug. 10, 2025. [Online]. Available: https://www.lyellcollection.org/doi/full/10.1144/geoenergy2024-002
[18] N. G. Holm, C. Oze, O. Mousis, J. H. Waite, and A. Guilbert-Lepoutre, 'Serpentinization and the Formation of H2 and CH4 on Celestial Bodies (Planets, Moons, Comets)', *Astrobiology*, vol. 15, no. 7, pp. 587–600, Jul. 2015, doi: 10.1089/ast.2014.1188.
[19] C. Zhang, '"This is getting exciting" | HyTerra's latest well yields hydrogen concentrations of 83%', hydrogeninsight.com. Accessed: Aug. 20, 2025. [Online]. Available: https://www.hydrogeninsight.com/production/this-is-getting-exciting-hyterras-latest-well-yields-hydrogen-concentrations-of-83-/2-1-1859083
[20] R. Roy *et al.*, 'Comparative techno-environmental analysis of grey, blue, green/yellow and pale-blue hydrogen production', *Int. J. Hydrog. Energy*, vol. 116, pp. 200–210, Apr. 2025, doi: 10.1016/j.ijhydene.2025.03.104.
[21] 'The colors of hydrogen – why green hydrogen matters - Stegra'. Accessed: Aug. 21, 2025. [Online]. Available:

https://stegra.com/news-and-stories/the-colors-of-hydrogen
[22] P. Martin, 'Green hydrogen "significantly more expensive to make" in the Netherlands than previously thought: government-funded study', hydrogeninsight.com. Accessed: Jun. 20, 2024. [Online]. Available: https://www.hydrogeninsight.com/production/green-hydrogen-significantly-more-expensive-to-make-in-the-netherlands-than-previously-thought-government-funded-study/2-1-1656096
[23] 'cost of hydrogen the netherlands green - Google Search'. Accessed: Jun. 15, 2024. [Online]. Available: https://www.google.com/search?q=cost+of+hydrogen+the+netherlands+green&rlz=1C1GCEA_enME1062ME1063&oq=cost+of+hydrogen+the+nether&gs_lcrp=EgZjaHJvbWUqBwgCECEYnwUyBggAEEUYOTIHCAEQIRigATIHCAIQIRifBdIBCDc1MjFqMG03qAIAsAIA&sourceid=chrome&ie=UTF-8
[24] W. I. Inc, 'Top Industrial Uses of Hydrogen | Industrial Hydrogen Safety', WHA International, Inc. Accessed: May 01, 2024. [Online]. Available: https://wha-international.com/hydrogen-in-industry/
[25] 'IPCC Assessment V6'. Accessed: Apr. 26, 2024. [Online]. Available: https://www.ipcc.ch/assessment-report/ar6/
[26] M. Barnard, 'How Much Does Hydrogen Leak And How Much Does It Matter?', CleanTechnica. Accessed: Aug. 09, 2025. [Online]. Available: https://cleantechnica.com/2024/10/21/how-much-does-hydrogen-leak-and-how-much-does-it-matter/
[27] M. K. BA, 'Pipeline Incidents Are a Daily Occurrence', FracTracker Alliance. Accessed: Aug. 09, 2025. [Online]. Available: https://www.fractracker.org/2025/04/pipeline-incidents-are-a-daily-occurrence/
[28] 'Rewiring Australia'. Accessed: Jun. 05, 2024. [Online]. Available: https://www.rewiringaustralia.org/
[29] 'Report Shows Pathway and Cost for Australia to Meet Climate Goals and Become Major Hydrogen Exporter', BloombergNEF. Accessed: Jun. 12, 2024. [Online]. Available: https://about.bnef.com/blog/report-shows-pathway-and-cost-for-australia-to-meet-climate-goals-and-become-major-hydrogen-exporter/
[30] 'Final modelling results | Net Zero Australia'. Accessed: May 01, 2024. [Online]. Available: https://www.netzeroaustralia.net.au/final-modelling-results/
[31] P. Martin *et al.*, 'A review of challenges with using the natural gas system for hydrogen', doi: 10.1002/ese3.1861.
[32] 'City cancels order of 50 hydrogen buses after realizing electric buses

make so much more sense | Electrek'. Accessed: Jun. 19, 2024. [Online]. Available: https://electrek.co/2022/01/11/city-cancels-order-50-hydrogen-buses-after-realizing-electric-buses-best/
[33] 'Exclusive: Poland's hydrogen buses confront prohibitive fuel costs, new report claims | Euronews'. Accessed: May 19, 2026. [Online]. Available: https://www.euronews.com/my-europe/2026/05/18/exclusive-polands-hydrogen-buses-confront-prohibitive-fuel-costs-new-report-claims
[34] H. Tosun, 'Are hydrogen fuel cell cars going to replace electric cars in the future? | LinkedIn', LinkedIn. Accessed: Apr. 30, 2024. [Online]. Available: https://www.linkedin.com/pulse/hydrogen-fuel-cell-cars-going-replace-electric-future-hamdi-tosun/
[35] 'Realistic roles for hydrogen in the future energy transition | Nature Reviews Clean Technology'. Accessed: May 17, 2026. [Online]. Available: https://www.nature.com/articles/s44359-025-00050-4
[36] 'A near 100 per cent renewables grid is well within reach, and with little storage | RenewEconomy'. Accessed: May 04, 2024. [Online]. Available: https://reneweconomy.com.au/a-near-100-per-cent-renewables-grid-is-well-within-reach-and-with-little-storage/
[37] 'Solar electricity every hour of every day is here and it changes everything | Ember'. Accessed: Aug. 09, 2025. [Online]. Available: https://ember-energy.org/latest-insights/solar-electricity-every-hour-of-every-day-is-here-and-it-changes-everything/chapter-3-getting-close-to-24-365-solar-generation/
[38] 'Our Work - h2eart.eu'. Accessed: May 17, 2024. [Online]. Available: https://h2eart.eu/our-work/
[39] O. of H. P. S. Australia, 'Office of Hydrogen Power South Australia', Office of Hydrogen Power South Australia. Accessed: Jun. 20, 2024. [Online]. Available: https://www.hydrogen.sa.gov.au
[40] 'U.S. National Clean Hydrogen Strategy and Roadmap | Hydrogen Program'. Accessed: Jun. 20, 2024. [Online]. Available: https://www.hydrogen.energy.gov/library/roadmaps-vision/clean-hydrogen-strategy-roadmap
[41] 'Transporting energy as hydrogen not electricity', Carbon Commentary. Accessed: Aug. 09, 2025. [Online]. Available: https://www.carboncommentary.com/blog/2022/1/20/transporting-energy-as-hydrogen-not-electricity
[42] M. M. Aba, I. L. Sauer, and N. B. Amado, 'Comparative review of hydrogen and electricity as energy carriers for the energy transition', *Int. J. Hydrog. Energy*, vol. 57, pp. 660–678, Feb. 2024, doi: 10.1016/j.ijhydene.2024.01.034.
[43] 'Green ammonia | Royal Society'. Accessed: May 01, 2024. [Online].

Available: https://royalsociety.org/news-resources/projects/low-carbon-energy-programme/green-ammonia/

[44] 'Executive Summary – Ammonia Technology Roadmap – Analysis', IEA. Accessed: Aug. 09, 2025. [Online]. Available: https://www.iea.org/reports/ammonia-technology-roadmap/executive-summary

[45] CSIRO, 'Food has a climate problem: Nitrous oxide emissions are accelerating with growing demand for fertilizer and meat – but there are solutions'. Accessed: Aug. 09, 2025. [Online]. Available: https://www.csiro.au/en/news/All/Articles/2024/June/global-nitrous-oxide-emissions

[46] Eye on AI, *Harnessing AI for Synthetic Biology | Anna Marie Wagner*, (Feb. 28, 2024). Accessed: Aug. 09, 2025. [Online Video]. Available: https://www.youtube.com/watch?v=-9HwbnCwi-w

[47] 'Pivot Bio Nitrogen Bacteria'. Accessed: Jul. 19, 2024. [Online]. Available: https://www.pivotbio.com/

[48] B. Gates, 'The Clean Industrial Revolution has arrived', gatesnotes.com. Accessed: Jul. 19, 2024. [Online]. Available: https://www.gatesnotes.com/Breakthrough-Energy-Summit-2024

[49] 'Hydrogen Demand Is Going To Fall Even Faster In Updated 2100 Projection - CleanTechnica'. Accessed: May 01, 2024. [Online]. Available: https://cleantechnica.com/2023/10/23/hydrogen-demand-is-going-to-fall-even-faster-in-updated-2100-projection/

[50] R. H. Mumm, P. D. Goldsmith, K. D. Rausch, and H. H. Stein, 'Land usage attributed to corn ethanol production in the United States: sensitivity to technological advances in corn grain yield, ethanol conversion, and co-product utilization', *Biotechnol. Biofuels*, vol. 7, p. 61, Apr. 2014, doi: 10.1186/1754-6834-7-61.

[51] Helton Bezzera, 'Hydrogen Report from PWC'. Accessed: May 16, 2024. [Online]. Available: https://www.linkedin.com/posts/helton-bezerra-84bb149_global-greenhydrogen-growth-activity-7189551848305938435-3Uyq/

[52] 'Debunking The Myth: Electrolyzing Hydrogen For Energy Is Wasteful & Higher Emissions - CleanTechnica'. Accessed: Mar. 19, 2025. [Online]. Available: https://cleantechnica.com/2025/03/11/debunking-the-myth-electrolyzing-hydrogen-for-energy-is-wasteful-higher-emissions/

[53] 'Pedal to the Metal 2024 - Global Energy Monitor'. Accessed: Aug. 12, 2025. [Online]. Available: https://globalenergymonitor.org/report/pedal-to-the-metal-2024/

[54] C. Coetzee, 'A bright future for UK steel'. Accessed: Aug. 09, 2025. [Online]. Available:

https://www.uselessgroup.org/publications/reports/bright-future-uk-steel

[55] B. Metal, 'Innovative metals processing', Boston Metal. Accessed: Jul. 19, 2024. [Online]. Available: https://www.bostonmetal.com/

[56] 'Green steelmaker Boston Metal to cut jobs following equipment failure', Canary Media. Accessed: May 18, 2026. [Online]. Available: https://www.canarymedia.com/articles/green-steel/startup-boston-metal-cuts-jobs-equipment-failure

[57] S. Kumar, Z. Xiong, J. Held, P. Bruggeman, and U. R. Kortshagen, 'Rapid carbon-free iron ore reduction using an atmospheric pressure hydrogen microwave plasma', *Chem. Eng. J.*, vol. 472, p. 145025, Sep. 2023, doi: 10.1016/j.cej.2023.145025.

[58] T. Leach and T. Horngren, 'Australia's main iron ore exports may not work with green steelmaking. Here's what we must do to prepare', The Conversation. Accessed: Aug. 10, 2025. [Online]. Available: http://theconversation.com/australias-main-iron-ore-exports-may-not-work-with-green-steelmaking-heres-what-we-must-do-to-prepare-201469

[59] G. Readfearn, 'Andrew Forrest says net zero is "fantasy" so his goal is "real zero". What does he mean – and can he achieve it?', *The Guardian*, Oct. 09, 2024. Accessed: Aug. 10, 2025. [Online]. Available: https://www.theguardian.com/environment/2024/oct/09/andrew-forrest-says-net-zero-is-fantasy-so-his-goal-is-real-zero-what-does-he-mean-and-can-he-achieve-it

[60] 'Uranium-235', *Wikipedia*. Apr. 28, 2026. Accessed: May 19, 2026. [Online]. Available: https://en.wikipedia.org/w/index.php?title=Uranium-235&oldid=1351526745

[61] Carmen, 'The world's ten largest uranium mines', Mining Technology. Accessed: Aug. 16, 2025. [Online]. Available: https://www.mining-technology.com/marketdata/ten-largest-uraniums-mines/

[62] 'Nuclear Power: Fission Or Fusion?', Ark Invest. Accessed: Jan. 30, 2025. [Online]. Available: https://www.ark-invest.com/articles/analyst-research/nuclear-power-fission-or-fusion

[63] 'Executive summary – Renewables 2023 – Analysis', IEA. Accessed: Jun. 02, 2024. [Online]. Available: https://www.iea.org/reports/renewables-2023/executive-summary

[64] 'Nuclear energy worldwide 2024 | GRS gGmbH'. Accessed: May 26, 2024. [Online]. Available: https://www.grs.de/en/news/nuclear-energy-worldwide-2024

[65] Ihédate and ihedate, 'World Nuclear Industry Status Report'. Accessed: Jun. 02, 2024. [Online]. Available: https://www.worldnuclearreport.org/
[66] 'Global Electricity Review 2026 | Ember'. Accessed: May 19, 2026. [Online]. Available: https://ember-energy.org/latest-insights/global-electricity-review-2026/
[67] 'Global Electricity Review 2025', Ember. Accessed: Jul. 24, 2025. [Online]. Available: https://ember-energy.org/latest-insights/global-electricity-review-2025
[68] 'Nuclear power capacity additions and retirements in selected countries and regions by decade in the Net Zero Scenario', IEA. Accessed: Jun. 03, 2024. [Online]. Available: https://www.iea.org/energy-system/electricity/nuclear-power
[69] 'Average gas and electricity usage | Ofgem'. Accessed: Aug. 17, 2025. [Online]. Available: https://www.ofgem.gov.uk/information-consumers/energy-advice-households/average-gas-and-electricity-use-explained
[70] 'Hinkley C: UK nuclear plant price tag could rocket by a third', Jan. 23, 2024. Accessed: Aug. 17, 2025. [Online]. Available: https://www.bbc.com/news/business-68073279
[71] 'Sizewell C cost "has doubled since 2020 and could near £40bn" | Sizewell C | The Guardian'. Accessed: Jan. 21, 2025. [Online]. Available: https://www.theguardian.com/business/2025/jan/14/sizewell-c-cost-nuclear-power-plant-edf
[72] 'Current Time - 2024 90 Seconds to midnight', Bulletin of the Atomic Scientists. Accessed: Jun. 14, 2024. [Online]. Available: https://thebulletin.org/doomsday-clock/current-time/
[73] 'US company claims nuclear battery breakthrough: Corporate - World Nuclear News'. Accessed: Jun. 14, 2024. [Online]. Available: https://world-nuclear-news.org/Articles/US-company-claims-nuclear-battery-breakthrough
[74] 'After Local Fault Test: Finnish Nuclear Power Plant Olkiluoto 3 Automatically Shuts Down Again - Renewable Energy Industry'. Accessed: Jun. 07, 2024. [Online]. Available: https://www.renewable-energy-industry.com/countries/article-6483-after-local-fault-test-finnish-nuclear-power-plant-olkiluoto-3-automatically-shuts-down-again
[75] 'SAEUL-2 - World Nuclear Association'. Accessed: Jun. 07, 2024. [Online]. Available: https://world-nuclear.org/nuclear-reactor-database/details/SAEUL-2
[76] D. Abbott, 'Limits to growth: Can nuclear power supply the world's needs?', *Bull. At. Sci.*, vol. 68, no. 5, pp. 23–32, Mar. 2012, doi:

10.1177/0096340212459124.
[77] G. M. Mudd, 'The Environmental sustainability of mining in Australia: key mega-trends and looming constraints', *Resour. Policy*, vol. 35, no. 2, pp. 98–115, Jun. 2010, doi: 10.1016/j.resourpol.2009.12.001.
[78] 'Nuclear Essentials - World Nuclear Association'. Accessed: May 08, 2024. [Online]. Available: https://www.world-nuclear.org/nuclear-essentials/how-is-uranium-made-into-nuclear-fuel
[79] 'Uranium Enrichment - World Nuclear Association'. Accessed: Aug. 13, 2025. [Online]. Available: https://world-nuclear.org/information-library/nuclear-fuel-cycle/conversion-enrichment-and-fabrication/uranium-enrichment
[80] 'Small Modular Reactors: Still too expensive, too slow and too risky'. Accessed: Jun. 02, 2024. [Online]. Available: https://ieefa.org/resources/small-modular-reactors-still-too-expensive-too-slow-and-too-risky
[81] CSIRO, 'GenCost: cost of building Australia's future electricity needs'. Accessed: May 28, 2024. [Online]. Available: https://www.csiro.au/en/research/technology-space/energy/GenCost
[82] 'Eye-popping new cost estimates released for NuScale small modular reactor'. Accessed: Jun. 07, 2024. [Online]. Available: https://ieefa.org/resources/eye-popping-new-cost-estimates-released-nuscale-small-modular-reactor
[83] J. Jolly, 'Rolls-Royce named winning bidder for UK small nuclear reactors', *The Guardian*, Jun. 10, 2025. Accessed: Aug. 19, 2025. [Online]. Available: https://www.theguardian.com/business/2025/jun/10/rolls-royce-uk-small-nuclear-reactors-smr-sizewell-c
[84] '"Micro" nuclear reactor start-up backed by Britain goes bust'. Accessed: Aug. 13, 2025. [Online]. Available: https://www.msn.com/en-us/news/technology/micro-nuclear-reactor-start-up-backed-by-britain-goes-bust/ar-AA1tcsc4
[85] 'Audioblog 15: The Pragmatic Climate Reset Part 1 — The Energy Transition Is Not Dead', Cleaning Up. Leadership in an Age of Climate Change. Accessed: Aug. 12, 2025. [Online]. Available: https://www.cleaningup.live/audioblog-15-the-pragmatic-climate-reset-part-1-the-energy-transition-is-not-dead/
[86] M. Barnard, 'Small nuclear reactor advocates refuse to learn the lessons of the past', The Future is Electric. Accessed: Aug. 11, 2025. [Online]. Available: https://medium.com/the-future-is-electric/small-nuclear-reactor-advocates-refuse-to-learn-the-

lessons-of-the-past-8ca1af3293c3

[87] 'Powered by Thorium'. Accessed: May 04, 2024. [Online]. Available: https://www.copenhagenatomics.com/

[88] 'TRIGA', *Wikipedia*. Aug. 05, 2024. Accessed: Sep. 26, 2024. [Online]. Available: https://en.wikipedia.org/w/index.php?title=TRIGA&oldid=1238806548#List_of_all_TRIGA_Nuclear_reactors_built_around_the_world

[89] 'Aalo - Welcome to the dawn of a Second Atomic Age'. Accessed: Sep. 26, 2024. [Online]. Available: https://www.aalo.com/

[90] 'List of fusion experiments', *Wikipedia*. Jul. 22, 2025. Accessed: Aug. 11, 2025. [Online]. Available: https://en.wikipedia.org/w/index.php?title=List_of_fusion_experiments&oldid=1301968504#Tokamak

[91] C. Seife, 'World's Largest Fusion Project Is in Big Trouble, New Documents Reveal', Scientific American. Accessed: Jan. 30, 2025. [Online]. Available: https://www.scientificamerican.com/article/worlds-largest-fusion-project-is-in-big-trouble-new-documents-reveal/

[92] 'ITER', *Wikipedia*. May 13, 2026. Accessed: May 20, 2026. [Online]. Available: https://en.wikipedia.org/w/index.php?title=ITER&oldid=1353933578

[93] B. Guarino, 'Will the World's First Nuclear Fusion Power Plant Be Built in Virginia? Here's Why We're Skeptical', Scientific American. Accessed: Jan. 30, 2025. [Online]. Available: https://www.scientificamerican.com/article/will-the-worlds-first-nuclear-fusion-power-plant-be-built-in-virginia-heres/

[94] 'Why Does Nuclear Power Plant Construction Cost So Much?', Institute for Progress. Accessed: Jun. 02, 2024. [Online]. Available: https://ifp.org/nuclear-power-plant-construction-costs/

[95] K. Dawson and P. Sabharwall, 'A Review of Light Water Reactor Costs and Cost Drivers', Idaho National Lab. (INL), Idaho Falls, ID (United States), INL/EXT-17-43273, Sep. 2017. doi: 10.2172/1466793.

[96] '2023 Levelized Cost Of Energy+', https://www.lazard.com. Accessed: May 05, 2024. [Online]. Available: https://www.lazard.com/research-insights/2023-levelized-cost-of-energyplus/

[97] 'Nuclear economics - lessons from Lazard to Hinkley Point-C | illuminem'. Accessed: May 29, 2024. [Online]. Available: https://illuminem.com/illuminemvoices/nuclear-economics-lessons-from-lazard-to-hinkley-point-c

[98] B. Flyvbjerg and D. Gardner, *How Big Things Get Done: The Surprising Factors That Determine the Fate of Every Project, from Home Renovations to Space Exploration and Everything in Between*. New York: Currency,

2023.
[99] 'Does Nuclear Energy Need the IRA Credits?', The Breakthrough Institute. Accessed: Aug. 14, 2025. [Online]. Available: https://thebreakthrough.org/issues/energy/does-nuclear-energy-need-the-ira-credits
[100] 'Nuclear power in France', *Wikipedia*. Aug. 11, 2025. Accessed: Aug. 14, 2025. [Online]. Available: https://en.wikipedia.org/w/index.php?title=Nuclear_power_in_France&oldid=1305389696#cite_note-1
[101] D. Nair, 'France Weighs Fresh Move to Nationalize $30 Billion Power Giant EDF', NDTV Profit. Accessed: Aug. 14, 2025. [Online]. Available: https://www.ndtvprofit.com/business/france-weighs-fresh-move-to-fix-30-billion-power-giant-edf
[102] 'France "far from ready" to build six new nuclear reactors, auditor says | Reuters'. Accessed: Jan. 21, 2025. [Online]. Available: https://www.reuters.com/business/energy/france-far-ready-build-six-new-nuclear-reactors-auditor-says-2025-01-14/
[103] E. Newburger, 'Biden grants PG&E $1.1 billion to keep Diablo Canyon nuclear plant open', CNBC. Accessed: May 26, 2024. [Online]. Available: https://www.cnbc.com/2022/11/21/biden-grants-pge-1-billion-to-keep-diablo-canyon-nuclear-plant-open.html
[104] 'White House holds summit on US nuclear energy deployment□: Nuclear Policies - World Nuclear News'. Accessed: Jun. 20, 2024. [Online]. Available: https://www.world-nuclear-news.org/Articles/White-House-holds-summit-on-US-nuclear-energy-depl
[105] 'Statement: $6 billion in subsidies for distressed nuclear plants are misguided', PennEnvironment. Accessed: Aug. 16, 2025. [Online]. Available: https://environmentamerica.org/pennsylvania/media-center/statement-6-billion-subsidies-distressed-nuclear-plants-are-misguided/
[106] 'FY 2026 Budget Justification', Energy.gov. Accessed: Aug. 16, 2025. [Online]. Available: https://www.energy.gov/cfo/articles/fy-2026-budget-justification
[107] O. E. Board, 'Ontario Energy Board issues decision on Ontario Power Generation's application for payment amounts'. Accessed: Aug. 11, 2025. [Online]. Available: https://www.newswire.ca/news-releases/ontario-energy-board-issues-decision-on-ontario-power-generation-s-application-for-payment-amounts-836711329.html
[108] M. Crawley, 'Why Doug Ford's government spends more than $6B/year subsidizing hydro rates', *CBC News*, Mar. 20, 2023. Accessed: Aug. 16, 2025. [Online]. Available:

https://www.cbc.ca/news/canada/toronto/ontario-budget-doug-ford-hydro-electricity-prices-1.6780846
[109] 'Canada's first SMR project: How is CAD20.9 billion cost calculated?', World Nuclear News. Accessed: Aug. 11, 2025. [Online]. Available: https://world-nuclear-news.org/articles/what-is-the-budget-for-canadas-first-smr-project
[110] 'China's quiet energy revolution: The switch from nuclear to renewable energy | RenewEconomy'. Accessed: Jun. 02, 2024. [Online]. Available: https://reneweconomy.com.au/chinas-quiet-energy-revolution-the-switch-from-nuclear-to-renewable-energy/
[111] 'Duke Energy Corporation - Second Quarter 2025 Earnings'. Accessed: Aug. 13, 2025. [Online]. Available: https://investors.duke-energy.com/events-and-presentations/events/event-details/2025/Second-Quarter-2025-Earnings/default.aspx
[112] 'How do you cool a reactor and does Australia have enough water for nuclear power?', *ABC News*, Jul. 19, 2024. Accessed: Aug. 16, 2025. [Online]. Available: https://www.abc.net.au/news/2024-07-20/nuclear-power-plant-water-supply-environmental-concerns-nsw/104084348
[113] T. Gaulkin, 'Counting the dead at Hiroshima and Nagasaki', Bulletin of the Atomic Scientists. Accessed: Aug. 14, 2025. [Online]. Available: https://thebulletin.org/2020/08/counting-the-dead-at-hiroshima-and-nagasaki/
[114] 'Isotopes of caesium', *Wikipedia*. Aug. 05, 2025. Accessed: Aug. 14, 2025. [Online]. Available: https://en.wikipedia.org/w/index.php?title=Isotopes_of_caesium&oldid=1304345230
[115] 'Finland built this tomb to store nuclear waste. Can it survive for 100,000 years?' Accessed: Aug. 14, 2025. [Online]. Available: https://www.science.org/content/article/finland-built-tomb-store-nuclear-waste-can-it-survive-100000-years
[116] 'Effects of the Chernobyl disaster', *Wikipedia*. Aug. 13, 2025. Accessed: Aug. 14, 2025. [Online]. Available: https://en.wikipedia.org/w/index.php?title=Effects_of_the_Chernobyl_disaster&oldid=1305614164
[117] 'Frequently Asked Chernobyl Questions'. Accessed: Feb. 20, 2025. [Online]. Available: https://www.iaea.org/newscenter/focus/chernobyl/faqs
[118] 'Cost to taxpayers to clean up nuclear waste jumps $100 billion in a year'. Accessed: Aug. 16, 2025. [Online]. Available: https://www.nbcnews.com/news/all/cost-taxpayers-clean-nuclear-waste-jumps-100-billion-year-n963586

[119] U. S. G. A. Office, 'Hanford Cleanup: Alternative Approaches Could Save Tens of Billions of Dollars | U.S. GAO'. Accessed: May 26, 2024. [Online]. Available: https://www.gao.gov/products/gao-23-106880
[120] 'Economics of Nuclear Power - World Nuclear Association'. Accessed: Jun. 03, 2024. [Online]. Available: https://world-nuclear.org/information-library/economic-aspects/economics-of-nuclear-power
[121] S. Laville, 'Push for new UK nuclear plants lacks facility for toxic waste, say experts', *The Guardian*, Mar. 28, 2022. Accessed: Aug. 14, 2025. [Online]. Available: https://www.theguardian.com/environment/2022/mar/28/push-for-new-uk-nuclear-plants-lacks-facility-for-toxic-waste-say-experts
[122] 'Nuclear Provision: the cost of cleaning up Britain's historic nuclear sites', GOV.UK. Accessed: May 26, 2024. [Online]. Available: https://www.gov.uk/government/publications/nuclear-provision-explaining-the-cost-of-cleaning-up-britains-nuclear-legacy/nuclear-provision-explaining-the-cost-of-cleaning-up-britains-nuclear-legacy
[123] 'Nuclear is likely to be key to the energy transition'. Accessed: Aug. 14, 2025. [Online]. Available: https://datawrapper.dwcdn.net/bFBR8/2/
[124] 'Nuclear and radiation accidents and incidents', *Wikipedia*. May 20, 2024. Accessed: Jun. 03, 2024. [Online]. Available: https://en.wikipedia.org/w/index.php?title=Nuclear_and_radiation_accidents_and_incidents&oldid=1224746918
[125] T. Rose and T. Sweeting, 'How safe is nuclear power? A statistical study suggests less than expected', *Bull. At. Sci.*, vol. 72, no. 2, pp. 112–115, Mar. 2016, doi: 10.1080/00963402.2016.1145910.
[126] 'Chernobyl disaster', *Wikipedia*. Jun. 02, 2024. Accessed: Jun. 03, 2024. [Online]. Available: https://en.wikipedia.org/w/index.php?title=Chernobyl_disaster&oldid=1226964539
[127] A. Brenner *et al.*, 'I-131 Dose Response for Incident Thyroid Cancers in Ukraine Related to the Chornobyl Accident', *Environmental Health Perspectives*, vol. 119. pp. 933–939, 2011. doi: 10.1289/ehp.1002674.
[128] P. Jacob *et al.*, 'Childhood exposure due to the Chernobyl accident and thyroid cancer risk in contaminated areas of Belarus and Russia', *British Journal of Cancer*, vol. 80. pp. 1461–1469, 1999. doi: 10.1038/sj.bjc.6690545.
[129] E. Cardis *et al.*, 'Risk of thyroid cancer after exposure to 131I in childhood.', *Journal of the National Cancer Institute*, vol. 97 10. pp. 724–32, 2005. doi: 10.1093/JNCI/DJI129.

[130] 'The Cancer Burden from Chernobyl in Europe WHO'. Accessed: Feb. 20, 2025. [Online]. Available: https://www.iarc.who.int/wp-content/uploads/2018/07/IARCBriefingChernobyl.pdf
[131] Y. Demidchik, V. Saenko, and S. Yamashita, 'Childhood thyroid cancer in Belarus, Russia, and Ukraine after Chernobyl and at present.', *Arquivos brasileiros de endocrinologia e metabologia*, vol. 51 5. pp. 748–62, 2007. doi: 10.1590/S0004-27302007000500012.
[132] 'Fukushima Daiichi nuclear disaster casualties', *Wikipedia*. May 14, 2024. Accessed: Jun. 03, 2024. [Online]. Available: https://en.wikipedia.org/w/index.php?title=Fukushima_Daiichi_nuclear_disaster_casualties&oldid=1223771242
[133] S. Yamashita, S. Suzuki, S. Suzuki, H. Shimura, and V. Saenko, 'Lessons from Fukushima: Latest Findings of Thyroid Cancer After the Fukushima Nuclear Power Plant Accident', *Thyroid*, vol. 28. pp. 11–22, 2018. doi: 10.1089/thy.2017.0283.
[134] A. Kumagai and K. Tanigawa, 'CURRENT STATUS OF THE FUKUSHIMA HEALTH MANAGEMENT SURVEY', *Radiation Protection Dosimetry*, vol. 182. pp. 31–39, 2018. doi: 10.1093/rpd/ncy138.
[135] E. Kim, O. Kurihara, N. Kunishima, T. Momose, T. Ishikawa, and M. Akashi, 'Internal thyroid doses to Fukushima residents—estimation and issues remaining', *Journal of Radiation Research*, vol. 57. pp. i118–i126, 2016. doi: 10.1093/jrr/rrw061.
[136] S. Nagataki, N. Takamura, K. Kamiya, and M. Akashi, 'Measurements of Individual Radiation Doses in Residents Living Around the Fukushima Nuclear Power Plant', vol. 180. pp. 439–447, 2013. doi: 10.1667/RR13351.1.
[137] S. Yasumura *et al.*, 'Achievements and Current Status of the Fukushima Health Management Survey', *Journal of Epidemiology*, vol. 32. pp. S3–S10, 2022. doi: 10.2188/jea.JE20210390.
[138] K. Kamiya, 'Health management and care following the Fukushima nuclear power plant accident: overview of Fukushima Health Management Survey', *Annals of the ICRP*, vol. 50. pp. 82–89, 2021. doi: 10.1177/01466453211015402.
[139] 'The American Cancer Society challenge goal to reduce US cancer mortality by 50% between 1990 and 2015: Results and reflections - Byers - 2016 - CA: A Cancer Journal for Clinicians - Wiley Online Library'. Accessed: Feb. 20, 2025. [Online]. Available: https://acsjournals.onlinelibrary.wiley.com/doi/full/10.3322/caac.21348
[140] 'Thyroid Cancer Survival Rates'. Accessed: Feb. 20, 2025. [Online]. Available: https://www.cancer.org/cancer/types/thyroid-cancer/detection-diagnosis-staging/survival-rates.html

[141] Kimiaki and Saito, 'Features of Exposure Doses to the Public due to the Fukushima Accident'. 2017.
[142] 'Three Mile Island accident', *Wikipedia*. May 22, 2024. Accessed: Jun. 03, 2024. [Online]. Available: https://en.wikipedia.org/w/index.php?title=Three_Mile_Island_accident&oldid=1225143671
[143] S. Wing, D. Richardson, D. Armstrong, and D. Crawford-Brown, 'A reevaluation of cancer incidence near the Three Mile Island nuclear plant: the collision of evidence and assumptions.', *Environ. Health Perspect.*, vol. 105, no. 1, pp. 52–57, Jan. 1997, doi: 10.1289/ehp.9710552.
[144] M. C. Hatch, S. Wallenstein, J. Beyea, J. W. Nieves, and M. Susser, 'Cancer rates after the Three Mile Island nuclear accident and proximity of residence to the plant.', *Am. J. Public Health*, vol. 81, no. 6, pp. 719–724, Jun. 1991, doi: 10.2105/ajph.81.6.719.
[145] A. George, 'Antarctica's "doomsday" glacier is heading for catastrophic collapse', New Scientist. Accessed: Aug. 17, 2025. [Online]. Available: https://www.newscientist.com/article/2448793-antarcticas-doomsday-glacier-is-heading-for-catastrophic-collapse/
[146] M. Barnard, 'Small nuclear reactor advocates refuse to learn the lessons of the past', The Future is Electric. Accessed: Aug. 11, 2025. [Online]. Available: https://medium.com/the-future-is-electric/small-nuclear-reactor-advocates-refuse-to-learn-the-lessons-of-the-past-8ca1af3293c3
[147] 'Lazard Releases 2025 Levelized Cost of Energy+ Report', https://www.lazard.com. Accessed: Aug. 17, 2025. [Online]. Available: https://www.lazard.com/news-announcements/lazard-releases-2025-levelized-cost-of-energyplus-report-pr/
[148] 'Global Status Report 2024', Global CCS Institute. Accessed: Aug. 29, 2025. [Online]. Available: https://www.globalccsinstitute.com/resources/global-status-report/
[149] 'Point Source Carbon Capture', netl.doe.gov. Accessed: Aug. 29, 2025. [Online]. Available: https://netl.doe.gov/carbon-management/carbon-capture
[150] M. Barnard, 'CCS Redux: "Best" Carbon Capture Facility In World Creates 25x More CO2 From Use Of Product', CleanTechnica. Accessed: Aug. 17, 2025. [Online]. Available: https://cleantechnica.com/2024/02/15/ccs-redux-best-carbon-capture-facility-in-world-creates-25x-more-co2-from-use-of-product/
[151] 'CCS Redux: "Best" Carbon Capture Facility In World Creates 25x More CO2 From Use Of Product - CleanTechnica'. Accessed: Jan. 18, 2025. [Online]. Available: https://cleantechnica.com/2024/02/15/ccs-

redux-best-carbon-capture-facility-in-world-creates-25x-more-co2-from-use-of-product/
[152] 'Why Carbon Capture and Storage Is Not a Net-Zero Solution for Canada's Oil and Gas Sector', International Institute for Sustainable Development. Accessed: Jan. 18, 2025. [Online]. Available: https://www.iisd.org/articles/deep-dive/carbon-capture-not-net-zero-solution
[153] M. Z. Jacobson, 'The health and climate impacts of carbon capture and direct air capture', *Energy Environ. Sci.*, vol. 12, no. 12, pp. 3567–3574, Dec. 2019, doi: 10.1039/C9EE02709B.
[154] S. Stapczynski, 'Big Oil's Climate Fix Is Running Out of Time to Prove Itself', *Bloomberg.com*. Accessed: Aug. 17, 2025. [Online]. Available: https://www.bloomberg.com/features/2023-carbon-capture-technology-running-out-of-time/
[155] 'Carbon-Sucking Tech Could Need More Energy Than All Homes Use - Bloomberg'. Accessed: May 21, 2026. [Online]. Available: https://www.bloomberg.com/news/articles/2023-03-21/carbon-sucking-tech-could-need-more-energy-than-all-homes-use
[156] 'U.S. unveils plans for large facilities to capture carbon directly from air'. Accessed: Jan. 18, 2025. [Online]. Available: https://www.science.org/content/article/us-unveils-plans-for-large-facilities-to-capture-carbon-directly-from-air
[157] V. Volcovici and V. Volcovici, 'US Energy Department restores funding to carbon removal projects', *Reuters*, Apr. 17, 2026. Accessed: May 21, 2026. [Online]. Available: https://www.reuters.com/legal/litigation/us-energy-department-restores-funding-carbon-removal-projects-2026-04-17/
[158] B. O. Þ. Alexandersson and V. Grettisson, 'Climeworks' capture fails to cover its own emissions', Heimildin. Accessed: May 16, 2025. [Online]. Available: https://heimildin.is/grein/24581/
[159] 'Prometheus Fuels', *Wikipedia*. Jul. 25, 2023. Accessed: Aug. 08, 2025. [Online]. Available: https://en.wikipedia.org/w/index.php?title=Prometheus_Fuels&oldid=1167054101
[160] 'Prometheus Slashes Cost of Carbon Capture by More Than 80%, Unlocking Affordable E-Fuels as a New Source of 24/7 Carbon Neutral Power'. Accessed: Aug. 07, 2025. [Online]. Available: https://finance.yahoo.com/news/prometheus-slashes-cost-carbon-capture-130000144.html
[161] P. Fuels, 'Prometheus Fuels announces breakthrough eKerosene pathway', BiobasedDieselDaily. Accessed: May 21, 2026. [Online]. Available: https://www.biobased-diesel.com/post/prometheus-

fuels-announces-breakthrough-ekerosene-pathway
[162] 'Novel coal-to-methanol process with near-zero carbon emission: Pulverized coal gasification-integrated green hydrogen process - ScienceDirect'. Accessed: May 17, 2024. [Online]. Available: https://www.sciencedirect.com/science/article/abs/pii/S0959652622001433
[163] 'The Role of E-fuels in Decarbonising Transport – Analysis', IEA. Accessed: Aug. 20, 2025. [Online]. Available: https://www.iea.org/reports/the-role-of-e-fuels-in-decarbonising-transport
[164] L. Collins, '"Expensive and polluting" | Why hydrogen-based e-fuels should not be considered a climate solution for road transport', hydrogeninsight.com. Accessed: Aug. 20, 2025. [Online]. Available: https://www.hydrogeninsight.com/transport/expensive-and-polluting-why-hydrogen-based-e-fuels-should-not-be-considered-a-climate-solution-for-road-transport/2-1-1450930
[165] 'E-fuels', T&E. Accessed: Aug. 20, 2025. [Online]. Available: https://www.transportenvironment.org/topics/cars/efuels
[166] 'Sustainable aviation fuel | Qantas Group'. Accessed: Aug. 21, 2025. [Online]. Available: https://www.qantas.com/au/en/qantas-group/sustainability/our-planet/sustainable-aviation-fuel.html
[167] D. Slanger, 'Unraveling Willingness to Pay for Sustainable Aviation Fuel', RMI. Accessed: Aug. 21, 2025. [Online]. Available: https://rmi.org/unraveling-willingness-to-pay-for-sustainable-aviation-fuel/
[168] 'We Are Eating the Earth: The Race to Fix Our Food System and Save Our Climate: Grunwald, Michael: 9781982160074: Amazon.com: Books'. Accessed: Jul. 20, 2025. [Online]. Available: https://www.amazon.com/We-Are-Eating-Earth-Climate/dp/1982160071/ref=sr_1_1
[169] T. J. Lark *et al.*, 'Environmental outcomes of the US Renewable Fuel Standard', *Proc. Natl. Acad. Sci. U. S. A.*, vol. 119, no. 9, p. e2101084119, Mar. 2022, doi: 10.1073/pnas.2101084119.
[170] 'Should Transportation Be Transitioned to Ethanol with Carbon Capture and Pipelines or Electricity? A Case Study | Environmental Science & Technology'. Accessed: Aug. 20, 2025. [Online]. Available: https://pubs.acs.org/doi/10.1021/acs.est.3c05054
[171] 'Green PE (Bio-PE) for extrusion, injection & blow molding - FKuR (EN)'. Accessed: Aug. 21, 2025. [Online]. Available: https://fkur.com/en/bioplastics/im-green-polyethylene/
[172] 'Ethanol Based Bioplastics', Greenfuels. Accessed: Aug. 21, 2025. [Online]. Available: https://www.green-fuels.org/focus/ethanol-

based-bioplastics
[173] P. M. R. https://www.polarismarketresearch.com, 'Biobased Polyvinyl Chloride Market Size, Share Global Analysis Report, 2024-2032', Polaris. Accessed: Aug. 21, 2025. [Online]. Available: https://www.polarismarketresearch.com/industry-analysis/biobased-polyvinyl-chloride-market
[174] P. Dziendzioł, S. Waśkiewicz, and K. Jaszcz, 'New Biobased Plasticizers for PVC Derived from Saturated Dimerized Fatty Acids', *Materials*, vol. 18, no. 9, p. 2155, May 2025, doi: 10.3390/ma18092155.
[175] 'Production of 1,3-Butadiene from Renewable Feedstocks Ethanol and 1,3-Butanediol - CRADA 560 (Final Report)'. Accessed: Aug. 21, 2025. [Online]. Available: https://www.pnnl.gov/publications/production-13-butadiene-renewable-feedstocks-ethanol-and-13-butanediol-crada-560-final
[176] L. Mou, J. Li, Y. Lu, G. Li, and J. Li, 'Polylactic acid: A future universal biobased polymer with multifunctional performance—from monomer synthesis, and processing to applications: A review', *J. Hazard. Mater. Adv.*, vol. 18, p. 100757, May 2025, doi: 10.1016/j.hazadv.2025.100757.
[177] 'Top ethanol producing countries worldwide 2024| Statista'. Accessed: Aug. 20, 2025. [Online]. Available: https://www.statista.com/statistics/281606/ethanol-production-in-selected-countries/
[178] 'Inside Elon Musk's Government Contracts', Built In. Accessed: Aug. 20, 2025. [Online]. Available: https://builtin.com/articles/elon-musk-government-contracts
[179] E. Howell, T. P. from D. Dobrijevic, and A. M. last updated, 'Starlink satellites: Facts, tracking and impact on astronomy', Space.com. Accessed: Apr. 30, 2024. [Online]. Available: https://www.space.com/spacex-starlink-satellites.html
[180] S. Insider, 'SpaceX Competitors and Similar Companies', Space Insider. Accessed: Aug. 20, 2025. [Online]. Available: https://spaceinsider.tech/2024/02/20/spacex-competitors-and-similar-companies/
[181] 'Chinese space program', *Wikipedia*. Aug. 19, 2025. Accessed: Aug. 20, 2025. [Online]. Available: https://en.wikipedia.org/w/index.php?title=Chinese_space_program&oldid=1306725149
[182] 'Diving Deep into Submarine Cables: The Undersea Lifelines of Internet Connectivity', Kentik. Accessed: Aug. 20, 2025. [Online]. Available: https://www.kentik.com/blog/diving-deep-into-submarine-cables-undersea-lifelines-of-internet-connectivity/

[183] 'Access to High Speed Internet is Key to Job Creation and Social Inclusion in the Arab World', World Bank. Accessed: Aug. 20, 2025. [Online]. Available: https://www.worldbank.org/en/news/press-release/2014/02/06/access-to-high-speed-internet-key-to-job-creation-social-inclusion-arab-world

[184] 'R&D as a percentage of GDP by country 2022', Statista. Accessed: May 05, 2024. [Online]. Available: https://www.statista.com/statistics/732269/worldwide-research-and-development-share-of-gdp-top-countries/

Made in the USA
Monee, IL
04 November 2026

67381721R00118